NUMBER THEORY
AND ITS HISTORY

Oystein Ore

DOVER PUBLICATIONS, INC., New York

Published in Canada by General Publishing Company, Ltd., 30 Lesmill Road, Don Mills, Toronto, Ontario.

Published in the United Kingdom by Constable and Company, Ltd.

This Dover edition, first published in 1988, is an unabridged republication of the work first published by the McGraw-Hill Book Company, Inc., New York, in 1948. Following page 358, it includes the *Supplement* which was added to later printings of the first edition.

Manufactured in the United States of America
Dover Publications, Inc., 31 East 2nd Street, Mineola, N.Y. 11501

Library of Congress Cataloging-in-Publication Data

Ore, Oystein, 1899–1968.
  Number theory and its history.

  Reprint. Originally published: New York : McGraw-Hill, 1948.
  Bibliography: p.
  Includes index.
  1. Numbers, Theory of. 2. Mathematics—History.
I. Title.
QA241.07 1988   512'.7   88-372
ISBN 0-486-65620-9 (pbk.)

PREFACE

This book is based upon a course dealing with the theory of numbers and its history which has been given at Yale for several years. Although the course has been attended primarily by college students in their junior and senior years it has been open to all interested. The lectures were intended to give the principal ideas and methods of number theory as well as their historical background and development through the centuries. Most texts on number theory contain inserted historical notes but in this course I have attempted to obtain a presentation of the results of the theory integrated more fully in the historical and cultural framework. Number theory seems particularly suited to this form of exposition, and in my experience it has contributed much to making the subject more informative as well as more palatable to the students.

Obviously, only some of the main problems of number theory could be included in this book. In making a selection, topics of systematic and historical importance capable of a simple presentation have been preferred. While many standard aspects of number theory had to be discussed, the treatment is often new, and much material has been added that has not heretofore made its appearance in texts. Also, in several instances I have found it desirable to introduce and define modern algebraic concepts whose usefulness is readily explained by the context.

The questions of number theory are of importance not only to mathematicians. Now, as in earlier days, these problems seem to possess a particular attraction for many laymen, and number theory is notable as one of the few fields of mathematics where the suggestions and conjectures of amateurs or nonprofessional mathematicians have exerted an appreciable influence. It may be mentioned incidentally that there have been few college classes that I can recall in which there were not to be found some students

v

who had already played with the strange properties of numbers. To make the theory available to readers whose mathematical knowledge may be limited, every effort has been made to reduce to a minimum the technical complications and mathematical requirements of the presentation. Thus, the book is of a more elementary character than many previous texts, and for the understanding of a greater part of the subject matter a knowledge of the simplest algebraic rules should be sufficient. Only in some of the later chapters has a more extended familiarity with mathematical manipulations been presupposed.

I am indebted to Prof. Otto Neugebauer for valuable comments on the historical material and to Paul T. Bateman for numerous suggestions for mathematical improvements that have been embodied in the text. In reading the proofs I was assisted by M. Gerstenhaber and E. V. Schenkman, who have also checked the numerical computations.

OYSTEIN ORE

NEW HAVEN, CONN.
August, 1948

CONTENTS

Chapter 9.   Congruences

Chapter 10.   Analysis of Congruences

Chapter 11.   Wilson's Theorem and Its Consequences

Chapter 12.   Euler's Theorem and Its Consequences

Chapter 13.   Theory of Decimal Expansions

Chapter 14.   The Converse of Fermat's Theorem

X   *CONTENTS*

Chapter 15.   The Classical Construction Problems

CHAPTER 1

COUNTING AND RECORDING OF NUMBERS

1-1. Numbers and counting. All the various forms of human culture and human society, even the most rudimentary types, seem to require some concept of *number* and some process for *counting*. According to the anthropologists, every people has some terminology for the first numbers, although in the most primitive tribes this may not extend beyond two or three. In a general way one can say that the process of counting consists in matching the objects to be counted with some familiar set of objects like fingers, toes, pebbles, sticks, notches, or the number words. It may be observed that the counting process often goes considerably beyond the existing terms for numerals in the language.

1-2. Basic number groups. Almost all people seem to have used their fingers as the most convenient and natural counters. In many languages this is easily recognized in the number terminology. In English we still use the term *digits* for the numerals. For numbers exceeding 10 the toes have quite commonly been used as further counters.

Very early in the cultural development it became necessary to perform more extensive counts to determine the number of cattle, of friends and foes, of days and years, and so on. To handle larger figures the counting process must be systematized. The first step in this direction consists in arranging the numbers into convenient *groups*. The choice of such basic groups depends naturally on the matching process used in counting.

The great preponderance of people use a basic *decimal* or *decadic* *group* of 10 objects, as one should expect from counting on the

1

fingers.   The word for 10 often signifies *one man.   Quinary systems* based on groups of 5 or *one hand* also occur, but the *vigesimal systems* based on a 20 group are much more common, corresponding of course to a complete count of fingers and toes.   Among the American Indian peoples the vigesimal system was in wide-spread use;   best known is the well-developed Mayan system. One finds traces of a 20 system in many other languages.   We still count in *scores*.   The French *quatre-vingt* for 80 is a remnant of a previously more extensive 20 count.   In Danish the 20 system is still used systematically for the names of numbers less than 100.

The largest known basic number, 60, is found in the *Babylonian sexagesimal system*.   It is difficult to explain the reasons for such a large unit group.   It has been suggested by several authors that it is the result of a merger of two different number systems.   We still use this system when measuring time and angles in minutes and seconds.   Other basic numbers than those mentioned here are quite rare.   We may detect a trace of a 12 or *duodecimal system* in our counts in dozens and gross.   Certain African tribes use basic groups of 3 and 4.   The *binary* or *dyadic system*, in which 2 or a pair is the basic concept, has been used in a rudimentary form by Australian indigenes.   The dyadic system is, however, a system whose simple properties often have a special mathematical use-fulness.

1–3. The number systems.   When the basic counting group is fixed, the numbers exceeding the first group would be obtained by counting afresh in a new group, then another, and so on.   For instance, in a quinary system where the basic five group might be called one $h(and)$, one would count one h. and one, one h. and two, $2h$. (10), $3h$. and 2 (17), and so on.   After one had reached five hands (25), one might say *hand of hands* (*h.h.*) and begin over again.   So as an example, one would denote 66 by $2hh$ and $3h$ and 1, that is, $2 \times 25 + 3 \times 5 + 1$.   Clearly this process can be extended indefinitely by introducing higher groups

$$hhh = 125 = 5^3 \qquad hhhh = 625 = 5^4$$

In this manner one arrives at a representation of any number as an expression

$$a_n \cdot 5^n + a_{n-1} \cdot 5^{n-1} + \cdots + a_2 \cdot 5^2 + a_1 \cdot 5 + a_0 \quad (1\text{--}1)$$

where each coefficient a_i is one of the numbers 0, 1, 2, 3, 4.

To be quite correct, one should observe that this particular example historically is fictitious, since no people is known to have developed and used a completely general system (1–1) with the base 5. But this systematic procedure for the construction of a number system was certainly the guiding principle in the evolution of our decadic number system and of many other systems. To confirm this assertion further one can turn to the philological analysis of our number terms. Through the laws of comparative linguistics one can trace a word like *eleven* to *one left over*, and similarly *twelve* to *two over*. There is some indication that our fundamental word *ten* may be derived from an Indo-European root meaning *two hands*. The word *hundred* comes from an original term *ten times* (*ten*). It is further interesting to note that the names for *thousand* are unrelated in the various main branches of the Indo-European languages; hence it is probably a rather late construction. The word itself seems to be derived from a Proto-Germanic term signifying *great hundred*.

In our decadic system all numbers are put in a form analogous to (1–1)

$$a_n \cdot 10^n + \cdots + a_2 \cdot 10^2 + a_1 \cdot 10 + a_0 \quad (1\text{--}2)$$

where the coefficients take values from 0 to 9. In general, in the subsequent chapters, we shall understand by a *number system with the base b* a system in which we represent the numbers in the form

$$a_n \cdot b^n + \cdots + a_2 \cdot b^2 + a_1 \cdot b + a_0 \quad (1\text{--}3)$$

where the coefficients a_i are numbers from 0 to $b - 1$.

It should be mentioned that relatively few peoples developed their number systems to this perfection. Also, in many languages one finds other methods for the construction of numbers. As an example of irregular construction let us mention that in Welsh the number words from 15 to 19 indicate 15, 15 + 1, 15 + 2, 2 × 9, 15 + 4. Subtraction occurs often as a method; for instance, in Latin, *un-de-viginti* = 20 − 1 = 19, *duo-de-sexaginta* = 60 − 2 = 58. Similar forms exist in Greek, Hindu, Mayan, and other languages.

The Mayan number system was developed to unusually high levels, but the system has one peculiar irregularity. The basic group is 20, but the group of second order is not 20 × 20 = 400 as one should expect, but 20 × 18 = 360. This appears to be connected with the division of the Mayan year into 18 months each consisting of 20 days, supplemented with 5 extra days. The higher groups in the system are

$$360 \times 20, \quad 360 \times 20^2, \cdots$$

1–4. Large numbers. As one looks at the development of number systems in retrospect it seems fairly simple to construct arbitrarily large numbers. However, in most systems the span of numbers actually used is very limited. Everyday life does not require very large numbers, and in many languages the number names do not go beyond thousands or even hundreds. We mentioned above that the term one thousand seems to have made a relatively late appearance in the Indo-European languages. The Greeks usually stopped at a *myriad* or ten thousand. For a long period the Romans did not have names or symbols for groups above 100,000. There exists in Rome an inscription on the Columna Rostrata commemorating the victory over Carthage at Mylae in the year 260 B.C. in which 31 symbols for 100,000 were repeated to signify 3,100,000. The Hindus had a peculiar attraction to large numbers, and immense figures occur commonly in their mythological tales and also in many of their algebraic problems. As a consequence, there existed particular names for the higher decadic groups to very great powers of 10. For instance, in a myth from the life of Buddha one finds the denominations up to 10^{153}.

Even our own number system has not been developed systematically to this extent. The word for one *million* is a fairly recent construction, which seems to have originated in Italy around A.D. 1400. The concept one *billion* has not found its final niche in our system. In American and sometimes in French terminology this means one thousand millions (10^9) while in most other countries of the world one billion is one million millions (10^{12}), while one thousand millions is called a *milliard*. It is probably only through the expenditures of the world wars that numbers of this size have reached such common use that confusion is likely to occur. When a *billion* is defined to be a thousand millions, a *trillion* becomes one thousand billions (10^{12}), a *quadrillion* one thousand trillions, and so on. On the other hand when a billion is one million millions, one million billions is a *trillion* (10^{18}), one million trillions is a *quadrillion* (10^{24}), and so on. While this discrepancy is not apt to cause any serious misunderstandings in everyday life, some universal agreement on usage and nomenclature would, nevertheless, be desirable.

The intellectual effort that lies behind a systematic extension of the number system is well illustrated by the fact that Archimedes (278–212 B.C.), the most advanced Greek mathematician, deems it worth while to devote a whole treatise, *The Sand Reckoning*, to this purpose. This work is addressed to his relative, King Gelon of Syracuse, and begins as follows:

There are some, King Gelon, who think that the number of grains of sand is infinite in multitude; and I mean by the sand not only that which exists about Syracuse and the rest of Sicily, but also that which is found in every region, whether inhabited or uninhabited. Again there are some, who, without regarding it as infinite, nevertheless think that no number has been named which is great enough to exceed its size.

Under this guise of aiming at finding a number exceeding the totality of grains of sand in the universe, as then known, Archimedes proceeds to construct a systematic enumeration method for arbitrarily high numbers.

1–5. Finger numbers. For the communication of numbers from one individual to another it is often desirable to have some other representation than the vocal expressions of the number names in the language. We now mainly use *written numbers*, a representation which we shall study subsequently. Before the advent of a fairly general writing ability the *finger numbers* were widely used as a universal numerical language. The numbers were indicated by means of different positions of fingers and hands. In a rudimentary way we still occasionally express numbers by our fingers. The finger numbers were in use in Europe both in the classical period and in the Middle Ages; they were used by the Greeks, Romans, Arabs, Hindus, and many other people. The human figures in ancient drawings and statues often show peculiar finger positions which denote numbers. For instance, Pliny states that the statue of Janus on the Forum in Rome represented the number 365, the days in the year, on its fingers.

In the Orient the finger numbers are still in common use. They enable buyers and sellers in the bazaars to bargain about prices independent of language differences. When the bargainers cover their hands with a piece of cloth, the finger numbers have the added advantage that the negotiations are secret to other parties.

Our best information about finger numbers in early times is due to the works of the Venerable Bede (A.D. 673–735), an English Benedictine monk from the cloisters in Wearmouth and Jarrow. His treatise *De temporum ratione* deals with the rules for calculating the date of Easter, and as an introduction it contains a description of the use of finger numbers (Fig. 1–1). The finger numbers were probably only in actual use for fairly moderate figures. Bede's numbers have a natural limit of 10,000, but he enlarges the method rather artificially so that it becomes possible to express numbers up to 1,000,000. To some limited extent it was possible to calculate with finger numbers. In Europe they seem to have disappeared gradually with the ascendency of the Hindu-Arabic number system.

1–6. Recordings of numbers. Neither the spoken numbers nor the finger numbers have any permanency. To preserve numbers for the purpose of records it is necessary to have other representations. Furthermore, without some memory aids the performance of calculations is extremely difficult.

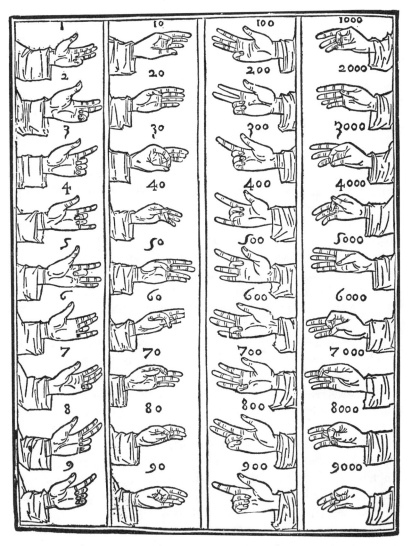

FIG. 1-1. Finger numbers. (*From Luca di Burgo Pacioli, Summa de arithmetica geometria, second edition, Venice, 1523. Courtesy of D. E. Smith Collection, Columbia University.*)

Many procedures have been devised to record numbers. The method of representing numbers by means of knots tied on strings has been used quite widely, in ancient China and on some of the South Sea Islands, and the *quipus* of the Incas in Peru are well known. In some localities split bamboo sticks have served as number records.

The most natural method for such records seems to consist in letting the counting process proceed by indicating each individual item through a mark on some suitable permanent material; for instance, dots or lines drawn in clay or on stones, scratches, notches, or scores on wooden sticks, chalk marks on slate or boards, and, of course, our present method of check marks on writing material.

The use of *wooden tallies* for recordings of numbers has been common in most European countries and in isolated districts it still occurs for special purposes. The English words *score* and *count*, from *computare* (*putare*, to cut), point to such methods. An important function of the tallies was to serve as contracts. In this case the tallies were ordinarily made in duplicate, one for each party, obtained by splitting a single piece of wood in two. Fraudulent changes were prevented quite effectively by cutting the number figures simultaneously over both parts. This system reached its highest level in the well-known Exchequer tallies, which formed an essential part of the British official accounting system from the twelfth century on (Fig. 1–2). On the Exchequer tallies the two pieces were unequal; the main piece, called the *stock*, served as a receipt while the separated, thinner leaf or *foil* was the record of payment. This tally system remained legally valid in England until the year 1826, and it had its official funeral pyre in 1834 when the burning of the accumulated tallies resulted in the fire that destroyed the old Parliament buildings.

1–7. Writing of numbers. The use of marks or notches to denote numbers is clearly a primitive form of writing, and it is likely to have been one of the first attempts in this direction. One can still see traces of this original procedure in many systems of number writing, for instance, quite plainly in the Roman numerals

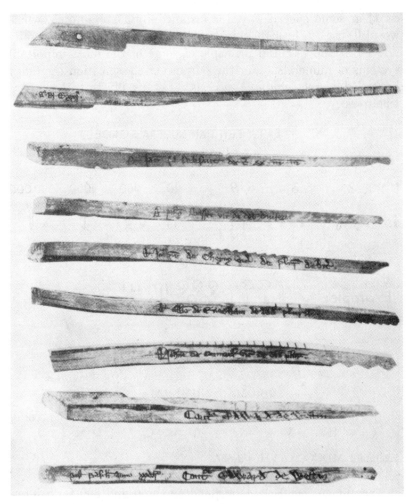

FIG. 1-2. Exchequer Tallies. (*From H. Jenkinson, Exchequer Tallies, Archae- logica, London,* 1911. *Courtesy of Society of Antiquaries, London.*)

I, II, III, IIII. To facilitate the reading of the tally marks when they became numerous, each basic group would naturally be indicated in some special way, for instance, by a cross-notch, which we still use. Through this simple procedure one has already arrived at the essential principle of some of the most important systems of numerals. For the purpose of classification they may be called *simple grouping systems*. Let us illustrate by a few examples:

EARLY EGYPTIAN NUMERALS (3400 B.C.)
(Hieroglyphic)

Quite familiar are the symbols of the Roman system:

ROMAN NUMERALS

1	2	5	10	50	100	500	1,000
I	II	V	X	L	C	D	M

Example: MDCCCXXVII = 1,827.

The Roman symbols corresponding to 50 and 500 form intermediate groupings within the basic decimal system, and they serve to clarify and simplify the writing of numbers. The subtraction principle in Roman numerals whereby a smaller unit preceding a higher one indicates subtraction (for instance, IX = 9, IV = 4), also shortens the representation. It may be mentioned that this use of subtraction in a systematic manner is an innovation of the last few centuries; in the classical period or even in medieval times it was used only rarely. Similar simplifications through subtractive notations occur in other numeral systems.

The Herodianic Greek numerals belong in principle to the simple grouping type:

ATTIC OR HERODIANIC GREEK NUMERALS

1	5	10	100	1000	10,000
I	Γ	△	H	X	M

Example: XΓHHH△△ΓII = 1827

These symbols are derived from the initials of the Greek numbers:

ΠΕΝΤΕ (5), ΔΕΚΑ (10), ΗΚΑΤΟΝ (100)

ΧΙΛΙΟΣ (1,000), ΜΥΡΙΟΣ (10,000)

The simple grouping system in several instances developed into a type of numeration that may be called a *multiplicative grouping system*. In such systems one has special ciphers for the numbers in the basic group, *e.g.*, 1, 2, . . . , 9, and a second class of symbols for the higher groups, *e.g.*,

$$10 = t, \qquad 100 = h, \qquad 1,000 = th, \cdots$$

The ciphers would then be used multiplicatively to show how many of the higher groups should be indicated. This would lead to representations of the type of the example

$$3,297 = 3th \quad 2h \quad 9t \quad 7$$

The traditional *Chinese-Japanese numeral system* is a multiplicative grouping as shown in the illustration. It should be noted that the writing is vertical instead of horizontal.

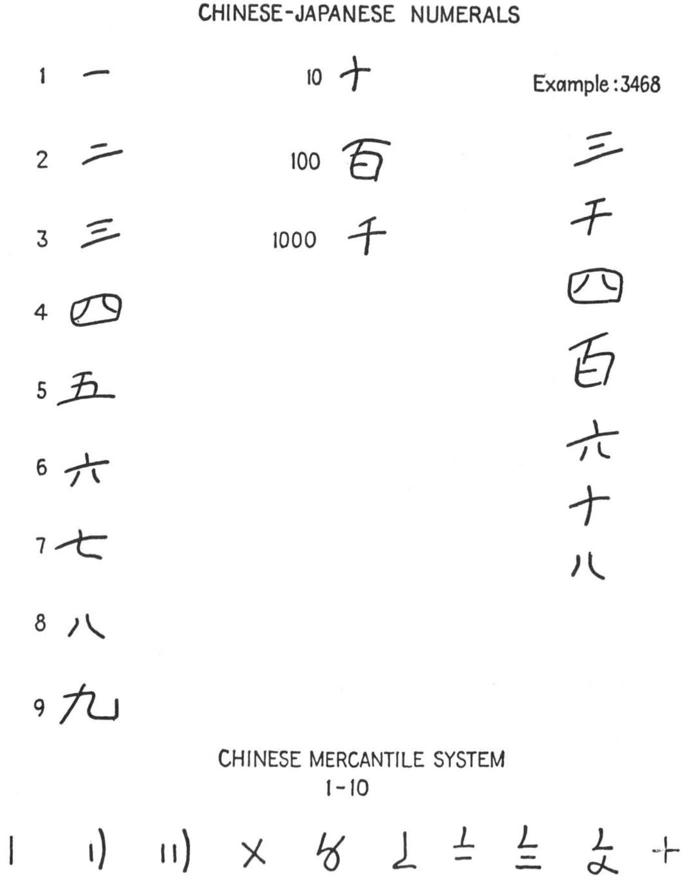

CHINESE-JAPANESE NUMERALS

CHINESE MERCANTILE SYSTEM
1-10

FIG. 1-3.

A third method of number writing may be called a *ciphered numeral system*. In the case of a decadic system one would denote the numbers from 1 to 9 by special symbols; similarly the multiples of 10 up to 90, the hundreds up to 900, and so on would have their individual signs. All numbers can then be represented as a combination of such symbols in a very compact form. The Egyptian hieroglyphic number writing later developed into the *hieratic* and *demotic* systems, which most nearly can be classified as being ciphered. Other examples are afforded by Coptic and Hindu Brahmi numerals.

HIERATIC NUMERALS

The usual Greek numerals are of a type that may be called *alphabetic*. The Greeks ciphered by means of the letters of the alphabet supplemented by a few symbols borrowed from the Semitic.

ALPHABETIC GREEK NUMERALS

1-9	α	β	γ	δ	ε	ς	ζ	η	θ
10-90	ι	κ	λ	μ	ν	ξ	o	π	φ
100-900	ρ	σ	τ	υ	φ	χ	ψ	ω	$\nearrow$

Example: $\psi \mu \beta = 742$

The higher units were obtained by special marks on the lower ones; for instance,

$$,\alpha = 1{,}000, \qquad ,\beta = 2{,}000$$

Alphabet numerals were used also by the Hebrews and the Syrians and in early Arabic and Gothic writing.

1–8. Calculations. Most of the numeral systems that we have mentioned in the preceding are not well suited for calculations, such as addition, subtraction, multiplication, and division. The peculiar difficulties that one encounters in performing these oper-

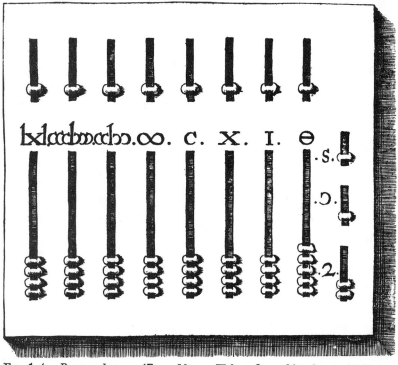

Fig. 1–4. *Roman abacus.* (*From Marcus Welser, Opera historica et philologica, Nürnberg, 1682*).

ations can easily be ascertained in the familiar system of Roman numerals. In most cultures the ability to handle computations has been considered an advanced and complicated art. On the other hand such knowledge is essential for the functioning of society when it reaches a certain stage of development; computations are essential for trade and commerce, for bookkeeping and accounting, and for many other purposes.

One finds as a consequence that devices to facilitate computations are in widespread use in the world. Best known is the *abacus* or *reckoning board*. The reckoning board was particularly an instrument of the merchants and tradesmen, and it could be applied universally regardless of differences in languages and numbers. This explains the close resemblance between Roman abaci, the Chinese *swanpan*, the Japanese *soroban*, the Russian *tschotu*. In the main they consist of balls in movable rows or on beads, not essentially different from the frames of balls used in our kindergartens to teach the rudiments of counting and calculation.

The only preserved Greek abacus was found on the island of Salamis. It is of a different type, a marble slab with engraved lines and Attic number symbols. There exist several Greek and Roman illustrations of persons using such abaci. The numbers were marked by means of small stones, in Latin *calculi*, whence the origin of our terms *calculus* and *to calculate*. In medieval Europe simplified abaci consisting only of lines, one each for units, tens, hundreds, and so on, were in common use. The abacus pattern could be drawn afresh on paper or parchment each time calculations were to be performed. The patterns could be carved permanently on a *comptoir* board or table, and they were often sewn on tablecloths, hence our word *bureau* derived from the Latin *burra* or woolen cloth. These checkered tabulating boards or abaci also originated such well-known terms as *exchequer* and *checks*. The process of calculating on these boards was called *casting* (on the lines), a term that is still preserved in various connections. We shall encounter it later in the ancient checking method for calculations known as casting out nines. The numbers were indicated on the board by means of special markers or counters, in French *jetons* (casters), or also by means of special reckoning coins stamped for this purpose. The abacus gradually lost ground as the knowledge of calculation with Hindu-Arabic numerals was spread. In modern times mechanized calculation has again gained the upper hand in any extended computations through the use of the calculating machine.

1–9. Positional numeral systems. We shall now turn to the history of our own numeral system in which we express every number by means of the digits 0, 1, 2, 3, 4, 5, 6, 7, 8, 9. It belongs to a type of numeral systems which are usually called *positional.* Such systems are based upon the principle of *local value,* so that a symbol designates a value or class which depends on the place it takes in the numeral representation. For instance, in the three numbers 352, 325, and 235, the digit 2 signifies respectively 2, 2 × 10, and 2 × 100.

Clearly the positional systems are closely related to the multiplicative grouping systems, and one obtains a positional system from a multiplicative grouping system simply by omitting the special symbols designating the higher class groups. As an example one may consider the Chinese-Japanese numerals. It may be observed, however, that this need not be the historical process through which a positional system originated.

The only complication which the positional notation involves lies in the necessity of introducing a *zero symbol* to express a void or missing class; for instance, 204 is different from 24. The essential discovery in the positional system may be considered to lie in the invention of this symbol. The many advantages of the positional system are not difficult to perceive. First, the numeral notation is very compact and easily readable. Next, it is possible to express arbitrarily large numbers only by the digits in the basic group. Finally, and not least important, in comparison with other systems the execution of calculations in the positional system becomes extremely simple.

The positional system is interesting culturally because it affords an illustration of an invention made independently in several civilizations. The earliest known numeral system to embody the principle of position is the sexagesimal Babylonian system, which we have mentioned previously. This system evolved from an earlier Sumerian system (about 3000 B.C.), which was also sexagesimal, but whose numeral representation was a simple grouping system. There exists an overwhelming material of Babylonian cuneiform tablets dated from 2000 B.C. to 200 B.C. that throws light upon the customs and institutions of this region. It is

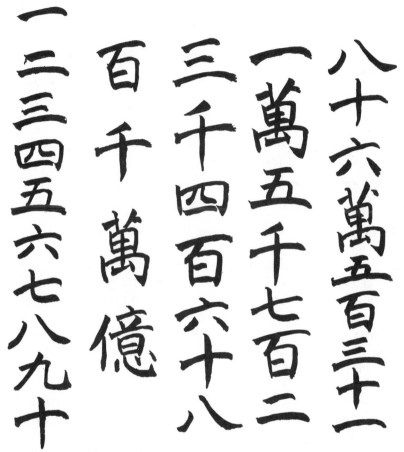

Fig. 1-5. Chinese-Japanese numerals (brush form). The first column gives the numbers 1–10, the second represents 100, 1,000, 10,000, 100,000,000. The three remaining columns give the examples 3,468, 15,702, and 860,531.

surprising that a considerable number of these tablets have been found to be mathematical texts and tables of a rather advanced nature. The numeral representations in the cuneiform texts use the symbols Γ and $\langle$ to denote 1 and 10 respectively. Within the basic 60 group the numbers are written by means of a simple grouping system, for instance,

$$= 35$$

To simplify the writing a subtractive symbol Γ , *lal* or minus, is applied, as in the example

$$= 20 - 1 = 19$$

The numbers exceeding 60 were written according to the positional principle. To illustrate,

$$= 1 \times 60^3 + 28 \times 60^2 + 52 \times 60 + 20 = 319,940$$

During a considerable part of the time in which they were in use, the Babylonian numerals were deficient because no sign for zero existed. As a consequence, the numeral representation was ambiguous. Often the true value of a number can be decided upon only through the context, although at times the spacing of the symbols may be of assistance. A zero sign does not come into regular use until after 300 B.C. Even so the numeral representation does not become unique since the zero is introduced only within the numeral and not at the end, so that, for instance, III may mean 3 or 3×60 or even 3×60^2.

The Mayans also achieved the distinction of having created a complete positional numeral system. Their number system, as we have already mentioned, was a vigesimal system with a deviation from the normal scheme in that the second number group was $360 = 20 \times 18$ instead of $400 = 20^2$. In the Mayan numeral system the first four numbers were denoted by dots, for instance, $\cdot\,\cdot\,\cdot = 3$. By crossing four dots one obtained the line ———, representing 5. The numbers in the basic 20 group were obtained from these two symbols by simple grouping, so that

$$\underline{\cdot\,\cdot} = 7, \qquad \underline{\overline{\cdot\,\cdot}} = 16$$

A symbol ⬭ for zero was used systematically as we do. Since the representation of the numerals was vertical, an example would appear as follows:

$$= 7 \times (18 \times 20^2) + 13 \times (18 \times 20) + 0 \times 20 + 10 = 55090$$

1-10. Hindu-Arabic numerals. Our numerals as we now use them are commonly known as the *Hindu-Arabic numerals*. Most historical evidence points to India as the country of their origin. To the Arabs who were instrumental in their transmission to Europe, they were known as the "Hindu numbers." Considerable material on early Hindu numerals is available from manuscripts and inscriptions. Although there is some difference of opinion among the scholars, it seems plausible that the number symbols from which our present digits have developed belonged to the Brahmi branch of numerals. This was originally a ciphered numeral system with the following first nine symbols:

BRAHMI SYMBOLS (100 B.C.)

1	2	3	4	5	6	7	8	9

The use of a positional system with a zero seems to have made its appearance in India in the period A.D. 600–800.

Around A.D. 800 the system was known among the Arabs in Bagdad and it gradually superseded the older type Arabic numerals. One of the greatest Arab mathematicians of this time was Mohammed ibn Musa al-Khowarizmi, whose work, *Al-Jabr wal-Muqabalah*, contributed much to the spread of calculations with the new system, first in the Arab world and later in Europe.

This treatise is of interest also because it is believed that its title *Al-Jabr* has given rise to the term *algebra* of modern mathematics.

Through the Arabs the Hindu numerals were introduced in Europe. An interesting early form, the *Gobar numerals*, appeared in Spain. The name *Gobar*, or *dust, numerals* is derived from the Indian custom of calculating on the ground or on a board covered with sand. The earliest preserved manuscript using Gobar numerals dates from A.D. 976. The Gobar numerals can also be found on the *apices* or *jetons* introduced by Gerbert, later Pope Sylvester II (died A.D. 1003), for calculations on the abacus.

GOBAR OR WESTERN ARABIC NUMERALS (1000 A.D.)

The works of al-Khowarizmi were translated into Latin, and through a perversion of his name the art of computing with Hindu-Arabic numerals became known as *algorism*. This term took on various other forms; in Chaucer it appears as *augrime*. The word is still preserved in mathematics, where a repeated calculating process is called an algorism. Other terms have been taken over from the Arabs. The Hindus early denoted the zero by a dot or a circle and used the term *śūnya*, or the *void*, for it. Translated into Arabic this became *as-sifi*, which is the common root of the words *zero* and *cipher*.

During the eleventh and twelfth centuries a number of European scholars went to Spain to study Arab learning. Among them one should mention the Englishmen Robert of Chester and Athelard of Bath, both of whom made translations of al-Khowarizmi's works. Still more important for the spread of the new numerals was the *Liber abaci* (A.D. 1202), a compendium of arithmetic, algebra, and number theory by Leonardo Fibonacci or Pisano, the only outstanding European mathematician of the Middle Ages. He expresses himself strongly in favor of calculations "modi Indorum," which he learned as a boy from Arab teachers in North Africa before returning to his native town of Pisa. Another text which was widely studied was the *Algorismus* of John of Halifax or Sacrobosco (about A.D. 1250).

Through the works of these and other scholars, but probably even more through merchants and trade, the knowledge of the Hindu-Arabic numerals was disseminated. The numerals took a great variety of shapes, some quite different from those now in use, but through the introduction of printing the forms became standardized and have since remained almost unchanged.

The transition to the new numerals was a long-drawn-out process. For several centuries there was considerable ill feeling between the algorismists, the users of the new numerals, and the abacists, who adhered to the abacus and the Roman numerals. Tradition long preserved Roman numerals in bookkeeping, coinage, and inscriptions. Not until the sixteenth century had the new numerals won a complete victory in schools and trade. Even as late as the famous work of Nikolaus Copernicus (died A.D. 1543) on the solar system, *De revolutionibus orbium coelestium*, one finds a strange mixture of Roman and Hindu-Arabic numerals and even numbers written out fully in words. The abacus or counter method of calculation remained in use much longer. To illustrate, let us quote from *The Ground of Artes* (1540) by Robert Recorde, one of the Englishmen who had most influence on arithmetic:

Both names are corruptly written: Arsemetrick for Arithmetic, as the Greeks call it, and Augrime for Algorisme as the Arabians found it; which both betoken the Science of Numbring, for Arithmos in Greek is called number: and of it comes Arithmetick, the Art of Numbring. So that Arithmetick is a Science or Art teaching the Manner and Use of Numbring: This Art may be wrought diversely, with Pen or with Counters. But I will first show you the working with the Pen, and then the other in order. [See Frontispiece.]

To complete this brief sketch of the development of our number system, it should be mentioned that the first satisfactory exposition of the use of decimal fractions was given by the Flemish

LIBRI.IIII.ARITHMETICE PRACTICAE

cõmunia ignores)tibi filiin medium ftatuerã ferre:quatenus his con
tentus ad altiores fecretiſſimaſɋ nũeroɮ poſthac fpeculationes trãſcẽ
das. DIS.Fideliter fœcifti prɇceptor colɇndiſſime:ſed tamẽ adhuc
unum ex promiſſis reftat declarandũ:quod forſitan tibi abduxit ob/
liuio. MAG.Q̃uod eft taleſ DIS.In.c.fecũdo thetoricɇ fpopõ/
deras præterhæc quɇ dicta funtquɇdam alia:& perpulchras artis cal
culatoriæ tradere regulas:quod(niſi tibi moleſtũ foret)te opere ad/
implere:& his finem arithmeticɇ tuɇimponere velim. MAG.Fa/
ciam ut petis.fcio equidem ɋ nihil ex pollicitis obliuiõe aut negligen
tia tranſis.Primo autẽ Algorithmũ vulgi p lineas:& denarios pro/
iectiles pandam:ut nõ modo per figuras numerorũ(quas cifrasvo
cant)veruẽtiã per denarios proiectiles quɇcuincɡ numerũ reprɇfenta/
re:addere:fubtrahere multiplicare aut diuidere infuper & numerũ ali
quẽ ignotũp notos reperire poſſis. Q̃uod quantũ & utilitatis & io /
cunditatis tibi allaturum fit:in fequentibus patebit. DIS. Eyaer/
go fermonem ad hæc quantocyus vertas.

Libri quarti Algorithmus cum denarijs proiectilibus: feu calcularis Tractatus quintus. Capitulum primũ. MAG. De Numeratione

2   D reprɇfentatiõnẽ numeri cũ denarijs piectilibuſ(quibus
p cifris utimur)neceſſariɇ funt lineɇ cifranũ rɇpfentãtes loca
eandemcɡ cum ipſis fignificatiõnẽ habentia ut infra.

Spacia vero fuß lineis contenta refpectu lineɇ ‚pximɇ fuprapofitɇ me
dietatẽ reprɇfentãt ut ♈   In linea ifta denarius pofitus/ decẽ fignifi/
cat:ſed is quiin fpacio cõfiftit/tm quincɡ reprɇfentat. Si igitur nume
rum aliquẽ p denarios proiectiles reprɇfentare volueris: tot denarios
fcdm linearũ &fpacioɮ exigentiã ad numerum ‚ppofitum reprɇfen/
tandum ponas:& cum.quincɡ denarios in linea aliqua habueris:pro
ipſis/leuatis fi placet in fpacio proximɇ fuperiori unũ ponas.Si ve/
ro in fpacio aliquo duos denarios reperies/ipſis leuatis/unũ ad lineã
immediate fuperiorẽ ponas.ut hic § Nec te fugiat p digiti applica/
tionem linearũ fignificatiõnẽ & augeri &minui poſſe.Ad quãcuncɡ

Fɪɢ. 1–6*a*. Instructions for computing on the lines in the form of a dialogue between magister and disciple. (*From Gregor Reisch, Margarita Philosophica, Strassbourg, 1504.*)

TRAC..V.ALGORiTHMI CALCVLARIS

enim lineam digitus applicatur:eadem unū tantam fignificabit: nec
tunc aliq̄ inferior: donec digitū depofueris aliqd repſentat.ut
☞ -o-o-o- ʒ.repreſentantq̄ depofito digito triginta important.Q uod
— ut ſingula facilius ac modo breuiori fiant in multiplicatione
& diuiſione huiuſcemodi digiti applicatione pſepe utendū eſt.

De Additione   Capitulum fecundum. DIS.

Dditionum(ni fallar ex dictis)eo modo fieri arbitror: ut ſcli
a   cet numerus cui debet fieri additio per denarios ad lineas cō ↄ
petentes ponatur:& ſimili modo numerus addendus ei appli
cetur. MAG. Optime ſentis ſi itidem exemplo probare potueris. -o-o-
DIS.Sit cauſa exempli numerus cui debet fieri additio.25ɔ. ita lo -o-o-
catus ↄ & numerus addendus.126.in hunc modum additus. -o-o-o-

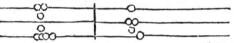

De Subtractione. Capitulum.iij.   MAG.

Et aliter ſubtractionē facies q̄ ut cum a quo debet fieri ſubtra
n   ctio p lineas competentes ponas:& ab eodē numerū ſubtra/
hendum ſubleuando tollas/in inferiori inchoādo.Si vero nu
merū aliquem a ſuperiori proximo ſubleuare nō poſſis:a linea proxi
me ſibi ſuprapoſita unū auferas/& illud in linea in qua ſubtractione;   π
perficere non potuiſti in decem reſoluens ſubtractionem ppoſitā per
ficies.Exemplū ſi a Ʂ .41.auferre voluero:unum quidē a linea infe
riori accipio & quattuor in linea ſecūda quero nec iuenio. ob hoc unū
de linea tertia accipiens/ipſum in linea inferiore in decē reſoluo qncꜧ
ſcilicet ponendo ad eandem lineam : unum vero ad ſpacium ſurſum
ut ſic π Illo facto de linea ſecunda tollantur .ſimiliter in ſimilibus
faciendum eſt.

De Multiplicatione Capitulum.iiij.' MAG.

I vero aliquem numerum multiplicare placuerit:ipſum ad li
ſ   neas competentes ponas:numerum autē multiplicātem mē/
te retineas:& p quolibet de nūero multiplicādo ſublato/mul/
tiplicāte integrum reſpectu lineę in qua operaris in loco dextro aut
ſiniſtro deponas:& idipſum uſcꜧ oēs multiplicādi mſtiplicati fuerint
ω iterandū eſt in ſupiore initiū ſumēdo.ut ſi cauſa exēpli ω .26.per.4
multiplicare voluens.26.ut pmiſſum eſt ad lineas ponani: atcꜧ in li
nea ſecūda inchoantes:unū demā:& p eodē in parte dextrā ad lineā
eandē.4.apponani.ut ſic φ & ſimili modo de ſecundo faciendū eſt.
ut hic ψ Et in eadē linea digito retento/eū qui in ſpacio inferiori pro
ximo reperiē accipiētes:ac p eodē(quia nō integerſed dimidius eſt)
nūintegra.4.ſed medietatē eius ſcꜧ.2.in lineā eandē ponam9. ut X

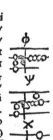

FIG. 1–6b. Examples of computation on the line.

mathematician Simon Stevin in his work *La Disme* (A.D. 1585). John Napier seems to have been one of the first to use the comma or point to separate decimals from the integers as we still do.

Bibliography

CAJORI, F.: *A History of Mathematical Notations*, The Open Court Publishing Company, Chicago, 1928.

———: *A History of Elementary Mathematics*, The Macmillan Company, New York, 1924.

HILL, G. E.: *The Development of Arabic Numerals in Europe Exhibited in Sixty-four Tables*, Clarendon Press, Oxford, 1915.

ROUSE-BALL, W. W.: *A Short Account of the History of Mathematics*, Macmillan & Co., Ltd., London, 1940.

SMITH, D. E.: *History of Mathematics*, Ginn & Company, Boston, 1923.

——— and L. C. KARPINSKI: *The Hindu-Arabic Numerals*, Ginn & Company, Boston, 1911.

CHAPTER 2

PROPERTIES OF NUMBERS. DIVISION

2-1. Number theory and numerology. The properties of the series of natural numbers, one of the basic and most essential concepts of mathematics, are the object of the theory of numbers. One finds that there exist many simple rules regarding numbers that are quite easy to discover and not too difficult to prove. However, number theory also includes an abundance of problems whose content can be comprehended and expressed in simple terms, yet whose solution has for centuries defied all mathematical investigation. Other problems whose solutions have been successfully obtained have yielded only to attacks by some of the most ingenious and advanced methods of modern mathematics. The simplicity in form of its problems and the great variation in the methods and tools for their solution explain the attraction that number theory has had for mathematicians and laymen. The innumerable individual contributions, calculations, speculations, and conjectures bear witness to the continued interest in this field of mathematics throughout the centuries.

The origins of the study of number properties go back probably almost as far as counting and the arithmetic operations. It does not take long before it is discovered that some numbers behave differently from others; for instance, some numbers can be divided into smaller equal parts and others not. The operations with fractions lead immediately to the study of divisibility of numbers, the least common multiple, and the greatest common divisor.

Other approaches have led to early number-theory questions. The solution of puzzles and amusement problems is one of them. To us who are accustomed to have an easy access to modern

25

diversions it may be difficult to realize the entertainment value that mathematical brain-teasers possessed for earlier generations. A brief revival of this interest may be seen in the great number of problems that circulated in the military camps during the war. This interest was attested by the numerous urgent requests for the correct answers, some from men earnestly interested (others undoubtedly put to settle bets). Many such problems are very old and appear in the earliest sources of mathematical information. The entertainment value of mathematical questions was developed especially by the Hindus. The Hindu mathematician Brahma-gupta (A.D. 588–660) states in one of his works: "These problems are stated merely for pleasure. The wise man can devise a thousand others or he can solve the problems of others by the rules given here. As the sun obscures the stars, so does the man of knowledge eclipse the glory of other mathematicians in an assembly of people by proposing algebraic problems and still more by solving them." According to tradition Bhaskara (about A.D. 1140) wrote his famous *Lilavati* ("the beautiful"), a collection of problems in poetic form, to comfort one of his daughters. The presentation of mathematical problems in verse was facilitated by the Hindu custom of using metaphors in the pronunciation of numbers.

A symbolic correspondence between numbers and objects or philosophical concepts and ideas was a trait common to many of the ancient cultures. One finds traces of such symbolism in most mythologies, and it is even preserved in some of our popular superstitions in regard to numbers. From this association, it is not a long step to speculations about properties of numbers and their implied relation to the corresponding concepts. Such *numerological* studies permeate the writings of the classical and medieval philosophers. The Pythagorean school (about 500 B.C.) was particularly devoted to symbolic number speculations in philosophy and nature. Their influence was considerable; even Plato touches upon numerology in several instances in his *Republic*, while Aristotle warns against arguments based upon such foundations. The later Neo-Pythagoreans tend to ascribe much of their mystical lore to the early school, but it seems certain that they have been under the influence also of other and sometimes even older sources. At present it is extremely difficult to ascribe any rational content to many of these numerological diatribes. To illustrate such passages let us make a few excerpts from a long eulogy on the number 7 from the work, *On the Creation of the World*, by the prominent Jewish philosopher Philo Judaeus:

And such great sanctity is there in the number seven, that it has a preeminent rank beyond all the other numbers in the first decade. For the other numbers, some produce without being produced, others are produced but have no productive power themselves; others again both produce and are produced. But the number seven alone is contemplated in no part. And this proposition we must confirm by demonstration. Now the number one produces all the other numbers in order, being itself produced absolutely by no other; and the number eight is produced by twice four, but itself produces no other number in the decade. Again, four has the rank of both, that is, of parents and offspring, for it produces eight when doubled, and it is produced by twice two. But seven alone, as I said before, neither produces nor is produced, on which account other philosophers liken this number to Victory, who has no mother, and to the virgin goddess, whom the fable asserts to have sprung from the head of Jupiter: and the Pythagoreans compare it to the Ruler of all things. . . .

Among the things then which are perceptible only by intellect, the number seven is proved to be the only thing free from motion and accident; but among things perceptible by the external senses, it displays a great and comprehensive power, contributing to the improvement of all terrestrial things and affecting even the periodical changes of the moon. And in what manner it does this, we must consider. The number seven when compounded of numbers beginning with the unit, makes eight-and-twenty, a perfect number, and one equalized in its parts.

Numerology has unquestionably stimulated investigations in number theory and bequeathed to us some most difficult problems. Let us mention the *perfect numbers*, which are equal to the sum of their aliquot parts (divisors). The discovery of a pair of *amicable numbers* symbolizing friendship, like 220 and 284, one the sum of the parts of the other, would not be possible without intimate study of divisibility properties of numbers.

A subject closely related to numerology is *gematria* (gematry), a name perhaps obtained as a corruption of the word *geometry*. By assigning number values to the letters in the alphabet in some order, each name and object received a number value. This letter weighting or gematry served to predict relations between persons or future events. Together with astrology it was one of the most popular ancient branches of superstitious learning, and both have persisted to our present days. The origin of gematry was directly connected with the form of the Hebrew and Greek numeral systems. Such alphabetic systems automatically assigned a number to each name and person. The names of the Bible have been a favorite field for gematry. Most famous is the Number of the Beast, given in the Revelation of St. John (13:18): "Here is wisdom. Let him that hath understanding count the number of the beast; for it is the number of a man and his number is six hundred three score and six." In spite of the innumerable researches on this question through the centuries it seems impossible to arrive at any definite solution. Clearly many names will have the same number. In the violent theological feuds of the Reformation it was a vicious stroke to write the opponent's name in such a way that his number became the fatal 666 of the beast. Let us mention also another number replacement that occurs in early theological writings. They often conclude with the number 99, in Greek $\varphi\theta$, and this is a gematry substitute for

$$\text{Amen} = \alpha\mu\eta\nu = 1 + 40 + 8 + 50 = 99$$

as one easily verifies by the list of Greek numerals.

2–2. Multiples and divisors. Number theory, as we have already stated, is primarily concerned with the properties of the natural numbers. However, it is convenient for most purposes to enlarge the system under consideration and investigate the whole set of integers

$$0, \pm 1, \pm 2, \pm \cdots \tag{2-1}$$

The two numbers $\pm a$ are sometimes said to be *associated*. They are characterized by the fact that they have the same absolute value $|a|$.

Now let a be an arbitrary integer. The *multiples* of a are all numbers

$$0, \pm a, \pm 2a, \pm \cdots \qquad (2\text{--}2)$$

i.e., all numbers of the form ka where k is integral. One sees that if ka and ha are two multiples of a, then their sum, difference, and product

$$ka \pm ha = (k \pm h)a, \qquad ka \cdot ha = kah \cdot a$$

are also multiples of a.

A simple example is the multiples $2n$ of 2, that is, the *even numbers*.

When a relation

$$c = ab \qquad (2\text{--}3)$$

holds between the integers a, b, and $c \neq 0$, one says that a is a *divisor* or *factor* of c and that c is *divisible* by a. We also call (2–3) a decomposition or factorization of c. Clearly b is also a divisor of c and uniquely determined by a. This leads to an observation that is useful in certain problems, namely, that the divisors of a number occur in pairs (a, b). The divisors in such a pair can only be equal (a, a) when $c = a^2$ is a square number.

From (2–3) one obtains a new factorization

$$c = (-a)(-b)$$

where the divisors are associated with a and b. Each number has the obvious decomposition

$$c = 1 \cdot c = (-1)(-c)$$

and ± 1 together with $\pm c$ are called *trivial* divisors. Other remarks about divisors are the following: If c_1 and c_2 are two numbers such that c_1 divides c_2 and conversely, then the two numbers are associated $c_1 = \pm c_2$. If $c_1 = ab_1$ and $c_2 = ab_2$ are two numbers divisible by a, then their sum and difference are divisible by a.

$$c_1 \pm c_2 = a(b_1 \pm b_2)$$

When $c = ab$ is divisible by a, and $d = cb_1$ is divisible by c, then $d = abb_1$ is divisible by a.

In most questions regarding divisors we shall assume tacitly that the number c is positive and that one only considers decompositions (2–3) with positive divisors a and b. Clearly all other factorizations can be written down as soon as the positive factorizations of positive integers have been obtained. In certain problems one is interested only in the *proper* divisors, consisting of all positive divisors including 1 that are actually less than c; that is, the number c is excluded. This is the point of view in the classical Greek problems.

In a decomposition (2–3) the factors a and b cannot both be greater than $\sqrt{c}$. One can suppose, therefore, that in a pair of divisors (a, b) one has $a \leq \sqrt{c}$ and $b \geq \sqrt{c}$. This limits the possible numbers that one has to try out in determining the factorizations of a number to divisors that do not exceed $\sqrt{c}$. For instance, when $c = 60$, one has $\sqrt{c} < 8$, and one finds the six pairs of divisors

1, 60	3, 20	5, 12
2, 30	4, 15	6, 10

Problems.

1. Find the divisors of the numbers 96 and 220.

2. Prove that a number is a square only when the number of (positive) divisors is odd.

2–3. Division and remainders. Let $b \neq 0$ be an arbitrary integer. Every other integer a will either be a multiple of b or fall between two consecutive multiples $q \cdot b$ and $(q + 1)b$ of b. Thus one can write

$$a = qb + r \tag{2-4}$$

where r is one of the numbers

$$0, 1, 2, \cdots, |b| - 1 \tag{2-5}$$

In (2–4) r is called the *least positive remainder* or simply the *remainder* of a by division with b, while q is the incomplete quotient or simply the *quotient*. As an example, let us divide 321 by 74

$$321 = 4 \cdot 74 + 25$$

Similarly, if 46 is divided by -17,

$$46 = (-2)(-17) + 12$$

It should be noted that when a and b in (2–4) are given, q and r are uniquely determined so that each integer a can be written in one way in the form $qb + r$, where r is one of the b numbers (2–5). For instance all numbers are *even* or *odd*, *i.e.*, belong to one of the two forms $2q$ or $2q + 1$. When these numbers are squared, one finds respectively

$$4q^2, \qquad 4q^2 + 4q + 1$$

so that we have:

THEOREM 2–1. The square of a number is either divisible by 4 or leaves the remainder 1 when divided by 4.

One can write the division (2–4) in the ordinary fractional form

$$\frac{a}{b} = q + \frac{r}{b}$$

where r/b is zero or a positive fraction less than 1 and q is the greatest integer that is less than or equal to a/b. Such quotients occur so often in number theory that it is convenient to introduce a special notation for them,

$$q = \left[\frac{a}{b}\right]$$

called *the greatest integer contained in* a/b.

Examples.

$$\left[\frac{27}{5}\right] = 5, \qquad \left[\frac{5}{3}\right] = 1, \qquad [2] = 2, \qquad \left[\frac{-1}{3}\right] = -1, \qquad \left[\frac{1}{3}\right] = 0$$

This notation may be extended to arbitrary real numbers. If for a real number

$$\alpha = q + \rho, \qquad 0 \leqq \rho < 1$$

then we write $q = [\alpha]$ for the integer q.

Examples.

$$[\pi] = 3, \qquad [e] = 2, \qquad \left[\frac{\pi^2}{2}\right] = 4$$

Sometimes there is an advantage in performing the division in a slightly different manner from (2–4): We select a multiple kb as near as possible to a on the number axis and obtain

$$a = kb + s \qquad (2\text{–}6)$$

where s now is a number between $-b/2$ and $b/2$. Such a representation as (2–6) we call a division with the *least absolute remainder*. Again k and s in (2–6) are uniquely determined except when b is even and the remainder is $s = \pm b/2$, when one can write the division in two ways

$$a = kb + \frac{b}{2} = (k+1)b - \frac{b}{2}$$

If it is desirable always to have a unique remainder one can agree to use $s = b/2$ in this case.

Example.

As examples of division with smallest absolute remainder, let us divide 35 by 9 and 46 by -17

$$35 = 4 \cdot 9 - 1, \qquad 46 = (-3)(-17) - 5$$

It is often convenient to apply the smallest absolute remainder in representing numbers. For instance, every number is representable in one of the three forms

$$3k, \qquad 3k \pm 1$$

or one of the five forms

$$5k, \qquad 5k \pm 1, \qquad 5k \pm 2$$

or in the four forms

$$4k, \qquad 4k + 2, \qquad 4k \pm 1$$

In the last classification the odd numbers must belong to the forms $m = 4k \pm 1$. As a consequence

$$m^2 = 16k^2 \pm 8k + 1 = 8k(2k \pm 1) + 1$$

so that we can say:

THEOREM 2-2. The square of an odd number is of the form $8q + 1$.

Examples.
$$5^2 = 3 \cdot 8 + 1, \qquad 7^2 = 6 \cdot 8 + 1$$

Some similar results are given among the problems.

Any number can be written in the form
$$n = 10a + b, \qquad 0 \leqq b \leqq 9$$

where b is the last digit in the decadic representation of the number. By squaring one obtains
$$n^2 = 100a^2 + 20ab + b^2$$

so that n^2 has the same last digit as b^2. But when one considers the squares of the numbers from 0 to 9, one finds that they end in one of the six digits 0, 1, 4, 5, 6, 9, so that one can say:

THEOREM 2-3. The last digit in the square of a number must be one of the numbers 0, 1, 4, 5, 6, 9.

For certain problems in number theory, for instance, with some factorization methods, it is of importance to be able to decide quickly whether a number can be a perfect square. By applying the same method as above to the last two digits of a number and looking up the squares of the numbers from 0 to 99, one finds that for a square the last two digits are limited to the following 22 possibilities:

Table of the last two digits in a square number

00	21	41	64	89
01	24	44	69	96
04	25	49	76	
09	29	56	81	
16	36	61	84	

Problems.

1. Divide the following pairs of numbers with respect to both least positive and least absolute remainders:

(a) 125 and 23   (b) 87 and 13

(c) −111 and −17   (d) 81 and 18

2. Prove that the square of a number not divisible by 2 or 3 is of the form $12n + 1$.

3. Prove that the fourth power of a number not divisible by 5 is of the form $5n + 1$.

4. Consider an analogue of theorem 2–3 for third and fourth powers.

5. Prove that in the decadic number system the fifth power of any number has the same last digit as the number itself.

6. Show that $n(n^2 - 1)$ is divisible by 24 when n is an odd number.

2–4. Number systems. As we mentioned previously, a variety of different number systems have been in use. We observed in this connection that when the basic counting group contained b elements, the systematic extension of the counting process would lead naturally to a representation of the natural numbers in the form

$$a = a_n \cdot b^n + a_{n-1} \cdot b^{n-1} + \cdots + a_2 \cdot b^2 + a_1 \cdot b + a_0 \quad (2\text{–}7)$$

where the numbers a_i take the values $0, 1, 2, \cdots, b - 1$.

In analogy to our numerals in the decimal system we can indicate the number (2–7) by the abbreviation

$$a = (a_n, a_{n-1}, \cdots a_2, a_1, a_0)_b$$

The question arises immediately how one can find the form of a number in a system with a given base number, or more generally how one can pass from one system to another. In (2–7) clearly the last number a_0 indicating the units is the least positive remainder of a by division with b,

$$a = q_1 \cdot b + a_0$$

where

$$q_1 = a_n \cdot b^{n-1} + \cdots + a_2 \cdot b + a_1$$

To determine a_1 one divides q_1 by b

$$q_1 = q_2 \cdot b + a_1$$

where

$$q_2 = a_n \cdot b^{n-2} + \cdots + a_2$$

When q_2 is divided by b one finds the remainder a_2, and through the repetition of this procedure all a_i's can be determined.

Examples.

1. To represent the number 1,749 in a system with the base **7**, one performs the divisions

$$1{,}749 = 249 \cdot 7 + 6$$
$$249 = 35 \cdot 7 + 4$$
$$35 = 5 \cdot 7 + 0$$

so that one finds

$$1{,}749 = (5, 0, 4, 6)_7$$

2. Similarly, to represent the number 19,151 to the base **12**:

$$19{,}151 = 1{,}595 \cdot 12 + 11$$
$$1{,}595 = 132 \cdot 12 + 11$$
$$132 = 11 \cdot 12 + 0$$

so that

$$19{,}151 = (11, 0, 11, 11)_{12}$$

In the preceding method for finding a number expressed to a base b, the digits $a_0, a_1, \ldots$ are determined from the lowest upward. One can also proceed in a manner that yields the digits in the reverse order $a_n, a_{n-1}, \ldots$. For this purpose, one determines the highest power of b such that b^n is less than a while the next power b^{n+1} exceeds a. Then from (2–7) it follows that the division of a by b^n must have the form

$$a = a_n \cdot b^n + r_{n-1}$$

where

$$r_{n-1} = a_{n-1} \cdot b^{n-1} + \cdots + a_0$$

From the remainder r_{n-1} one determines a_{n-1} in the same manner, and so on. This method is facilitated by a table of the various powers of the base number b.

Example.

Represent 1,832 to the base **7**. One calculates

$$7^2 = 49. \qquad 7^3 = 343. \qquad 7^4 = 2.401$$

and from the divisions

$$1{,}832 = 343 \cdot 5 + 117$$
$$117 = 49 \cdot 2 + 19$$
$$19 = 7 \cdot 2 + 5$$

one concludes

$$1{,}832 = (5, 2, 2, 5)_7$$

From time to time it has been suggested that our venerable decimal system be discarded in favor of some other system. Most often the numbers 6, 8, or 12 are proposed as the new bases. The arguments for such a change are of various kinds. In the case of the bases 6 and 12, it is pointed out that division by 3 becomes simple; in decimals one has the infinite expansion

$$\tfrac{1}{3} = 0.333 \cdots$$

while with the bases 6 or 12

$$\tfrac{1}{3} = (0, 2)_6 = (0, 4)_{12}$$

On the other hand, fractions with denominator 5 would become complicated in these systems; for instance,

$$\tfrac{1}{5} = (0, 1, 1, \cdots)_6$$

If one should wish to have simple expansions for all fractions with denominators 2, 3, 4, 5, one would be led to the Babylonian sexagesimal system. Large bases will give short representations of numbers, but they have the drawback that the size of the multiplication tables to be memorized is considerably increased. A 12×12 multiplication table instead of the usual 10×10 table may be admissible, but a 60×60 table is clearly out of the question. Small bases lead to long number representations but very simple multiplication tables. On the whole there is little evidence that a change of bases will materially reduce the time consumed by numerical computations. The reformers usually pass lightly over the resulting complications and the necessity of changing records, tables, and machines. For one thing, in order to avoid a state of utter confusion in the transition period it would be necessary to invent and use a completely new system of ciphers, because otherwise no one would know whether 23 should mean 23 or 19 (if the base were 8) or 27 (if the base 12 had been decided upon).

Problems.

1. Write the two numbers 1,947 and 21,648 to the four bases 3, 5, 7, and 23.
2. Write the number of seconds in 24 hours in the sexagesimal system.
3. Write the number of seconds of arc in 360° in the sexagesimal system.

2–5. Binary number systems. Number systems with other bases than 10 have applications in several branches of mathematics; particularly, the use of low base numbers 2 and 3 is helpful in many types of problems. In the *triadic* or *ternary system* each number is represented by means of the digits 0, 1, and 2 while in the *dyadic* or *binary system* each number appears as a series of marks 0 or 1. As an example, let us expand the number 87 to the base 2. One finds as before

$$87 = 43 \cdot 2 + 1$$
$$43 = 21 \cdot 2 + 1$$
$$21 = 10 \cdot 2 + 1$$
$$10 = 5 \cdot 2 + 0$$
$$5 = 2 \cdot 2 + 1$$
$$2 = 1 \cdot 2 + 0$$
$$1 = 0 \cdot 2 + 1$$

Hence

$$87 = (1, 0, 1, 0, 1, 1, 1)_2 = 2^6 + 2^4 + 2^2 + 2^1 + 2^0 \quad (2\text{-}8)$$

Let us consider the series of numbers

$$87, 43, 21, 10, 5, 2, 1 \quad (2\text{-}9)$$

occurring in the divisions. Each number is half the preceding with the remainder thrown away. One obtains the digits in (2-8), in reverse order, by writing 0 or 1 for each number in (2–9) depending on whether it is even or odd. This schematic method not only simplifies the determination of the representation in the dyadic system, but leads also to a peculiar multiplication procedure. To illustrate let us multiply 87 by 59. We form two chains of numbers, the first obtained by successively taking half the preceding number as above, the second proceeding by doubling.

87	59
43	118
21	236
10	~~472~~
5	944
2	~~1,888~~
1	3,776
	5,133

In the second column one strikes out the numbers corresponding to even numbers in the first, and the sum of the remaining terms gives the desired product. The proof lies in the dyadic representation of 87

$$87 \cdot 59 = (1 + 2 + 2^2 + 2^4 + 2^6)59$$

This method for performing multiplication reduces the operations to addition, together with doubling or duplication, and halving or mediation. In medieval treatises on computation, these two processes of duplication and mediation were considered to be separate arithmetic operations besides the four usual ones. The principle of reducing multiplication to duplication is very old; it was used by the early Egyptians, and it may well have been the first approach to a systematic multiplication procedure. The method given above is sometimes called *Russian multiplication* because of its use among Russian peasants. Its great advantage to the inexperienced calculator lies in the fact that it makes unnecessary the memorizing of the multiplication table.

There are a great number of games and puzzles whose solutions depend on the use of the dyadic number system. One is the fairly well-known Chinese game of *Nim*, which is discussed at some length in two of the books cited below (Hardy and Wright, and Uspensky and Heaslet). Another puzzle for children consists of a set of cards, each with a certain group of numbers on it. One is asked to think of a number, and to indicate on which cards it can be found. It is then possible immediately to pronounce the number in question. The cards contain the numbers up to a certain limit arranged in such a way that the first card contains all numbers whose lowest digit in the dyadic system is 1, that is, the odd numbers; the second contains all numbers whose second digit is 1, beginning with 2; the third all whose third digit is 1, beginning with 4, and so on. When it is known on which cards a given number occurs, its dyadic expansion is known. The number itself is the sum of the first numbers on the cards where it appears. As a simple example let us take four cards containing all numbers less than $2^4 = 16$. One finds that they must have the forms

1	9	2	10	4	12	8	12
3	11	3	11	5	13	9	13
5	13	6	14	6	14	10	14
7	15	7	15	7	15	11	15

Problems.

1. Construct such cards for all numbers up to 31.
2. Expand 365 to the bases 2 and 3.
3. Multiply 178 and 147 by Russian multiplication.

Bibliography

ALBERT, A. A.: *College Algebra*, McGraw-Hill Book Company, Inc., New York, 1946.

BELL, E. T.: *Numerology*, The Williams & Wilkins Company, Baltimore, 1933.

————: *The Magic of Numbers*, McGraw-Hill Book Company, Inc., New York, 1946.

BIRKHOFF, G., and S. MACLANE: *A Survey of Modern Algebra*, The Macmillan Company, New York, 1941.

HARDY, G. H., and E. M. WRIGHT: *An Introduction to the Theory of Numbers*, Clarendon Press, Oxford, 1938.

HOPPER, VINCENT F.: *Medieval Number Symbolism*, Columbia University Press, New York, 1938.

MACDUFFEE, C. C.: *An Introduction to Abstract Algebra*, John Wiley & Sons, Inc., New York, 1940.

USPENSKY, J. V., and M. A. HEASLET: *Elementary Number Theory*, McGraw-Hill Book Company, Inc., New York, 1939.

CHAPTER 3

EUCLID'S ALGORISM

3–1. Greatest common divisor. Euclid's algorism. Let a and b be two integers. If a number c divides a and b simultaneously, we shall call it a *common divisor* of a and b. Among the common divisors of two numbers there must exist a greatest one, which we shall call the *greatest common divisor* (g.c.d.) of a and b. It is usually denoted by the symbol (a, b). Since every number has the divisor 1 it follows that (a, b) is a positive number. If $(a, b) = 1$ we say that the two numbers are *relatively prime*. In this case ± 1 are the only common divisors.

Examples.

When the divisors of the two numbers 24 and 56 are determined one finds that their g.c.d. is 8. The numbers 15 and 22 are relatively prime.

We shall now prove:

THEOREM 3–1. Any common divisor of two numbers divides their greatest common divisor.

To establish this theorem we shall introduce a procedure known as *Euclid's algorism,* one of the basic methods of elementary number theory. It occurs in the seventh book of Euclid's *Elements* (about 300 B.C.); however it is certainly of earlier origin. Let a and b be the two given numbers whose g.c.d. is to be studied. Since there is only question of divisibility, there is no limitation in assuming that a and b are positive and $a \geq b$. We divide a by b with respect to the least positive remainder

$$a = q_1 b + r_1, \qquad 0 \leq r_1 < b$$

Next we divide b by r_1

$$b = q_2 r_1 + r_2, \qquad 0 \leq r_2 < r_1$$

and continue this process on r_1 and r_2, and so on. Since the remainders r_1, r_2, ... form a decreasing sequence of positive integers, one must finally arrive at a division for which $r_{n+1} = 0$

$$\left.\begin{aligned}
a &= q_1 b + r_1 \\
b &= q_2 r_1 + r_2 \\
r_1 &= q_3 r_2 + r_3 \\
&\cdots\cdots\cdots \\
r_{n-2} &= q_n r_{n-1} + r_n \\
r_{n-1} &= q_{n+1} r_n
\end{aligned}\right\} \tag{3-1}$$

Example.

Let us perform Euclid's algorism on the two numbers 76,084 and 63,020.

$$\begin{aligned}
76{,}084 &= 63{,}020 \cdot 1 + 13{,}064 \\
63{,}020 &= 13{,}064 \cdot 4 + 10{,}764 \\
13{,}064 &= 10{,}764 \cdot 1 + 2{,}300 \\
10{,}764 &= 2{,}300 \cdot 4 + 1{,}564 \\
2{,}300 &= 1{,}564 \cdot 1 + 736 \\
1{,}564 &= 736 \cdot 2 + 92 \\
736 &= 92 \cdot 8
\end{aligned}$$

We shall now show that in Euclid's algorism (3–1) the last nonvanishing remainder r_n is the g.c.d. of a and b. The first step is to show that r_n divides a and b. It follows from the last division in (3–1) that r_n divides r_{n-1}. The next to the last division shows that r_n divides r_{n-2} since it divides both terms on the right. Similarly from

$$r_{n-3} = q_{n-1} r_{n-2} + r_{n-1}$$

one concludes that r_n divides r_{n-3}, and successively one sees that r_n divides all r_i's and finally a and b.

The second step consists in showing that every divisor c of a and b divides r_n; this clearly implies that r_n is the g.c.d. of a and b

and has the property required by theorem 3-1. But from the first division in (3-1) one sees that any common divisor c of a and b divides r_1, since $r_1 = a - q_1b$; from the second, in the same way, c divides r_2; and by continuing this process one establishes that all r_i's, and hence r_n, are divisible by c.

Example.

From the previous algorism on the two numbers 76,084 and 63,020 it follows that their g.c.d. is 92.

Euclid's algorism gives a very simple and efficient method for the determination of the g.c.d. of two numbers. The French mathematician Lamé (1795–1870) has shown that the number of divisions in the algorism is at most five times the number of digits in the smaller number. Another observation of importance is that all arguments used above remain valid for any chain of relations (3-1) without any limitations on the numbers r_i; therefore, the conclusions are the same, except that r_n may possibly be the negative value of the g.c.d. One could, for instance, have used the least absolute remainders in the divisions (3-1). It has been shown by the German mathematician Kronecker (1823–1891), one of the leading contributors to number theory in the last century, that no Euclid algorism can be shorter than the one obtained by least absolute remainders. (For a more detailed study of the algorism, see the book by Uspensky and Heaslet, cited in the bibliography of Chap. 2.)

Example.

Let us perform the algorism for 76,084 and 63,020 by least absolute remainders.

$$76,084 = 63,020 \cdot 1 + 13,064$$
$$63,020 = 13,064 \cdot 5 - 2,300$$
$$13,064 = 2,300 \cdot 6 - 736$$
$$2,300 = 736 \cdot 3 + 92$$
$$736 = 92 \cdot 8$$

Problems.

Find Euclid's algorism for least positive and least absolute remainders and determine the g.c.d. for the pairs of numbers:

1. 139 and 49 **2.** 1,124 and 1,472 **3.** 17,296 and 18,416

3-2. The division lemma. From the algorism of Euclid one can derive various other properties of the g.c.d. An important consequence is the *division lemma:*

THEOREM 3-2. When a product ac is divisible by a number b that is relatively prime to a, the factor c must be divisible by b.

Proof: Since a and b are relatively prime, the last remainder r_n in the algorism must be 1 so that it has the form

$$a = q_1 b + r_1$$

$$\cdot \quad \cdot \quad \cdot \quad \cdot \quad \cdot \quad \cdot \quad \cdot$$

$$r_{n-2} = q_n r_{n-1} + 1$$

We multiply each of these equations by c and obtain

$$ac = q_1 bc + r_1 c$$

$$\cdot \quad \cdot \quad \cdot \quad \cdot \quad \cdot \quad \cdot \quad \cdot$$

$$r_{n-2} c = q_n r_{n-1} c + c$$

Since ac is divisible by b according to our assumption, the first relation shows that $r_1 c$ is divisible by b. From the second relation one finds that $r_2 c$ is divisible by b, and successively one finds that all $r_i c$ and finally c are divisible by b, as we set out to show.

Theorem 3-2 leads to the further result:

THEOREM 3-3. When a number is relatively prime to each of several numbers, it is relatively prime to their product.

Proof: Let a be relatively prime to b and to c. If a has a common divisor d with bc, the product is divisible by d. But $(d, b) = 1$ since d divides a; thus d must divide c, according to theorem 3-2, contrary to the fact that also $(d, c) = 1$. The extension of theorem 3-3 to several factors is immediate.

Another consequence of Euclid's algorism is:

THEOREM 3–4. For the greatest common divisor of two products ma and mb, one has the rule

$$(ma, mb) = m(a, b) \qquad (3\text{--}2)$$

Proof: In Euclid's algorism (3–1) for the numbers a and b let us multiply each equation by m.

$$am = q_1 bm + r_1 m$$
$$\cdot \quad \cdot \quad \cdot \quad \cdot \quad \cdot \quad \cdot \quad \cdot \quad \cdot$$
$$r_{n-2}m = q_n r_{n-1}m + r_n m$$
$$r_{n-1}m = q_{n+1} r_n m$$

Clearly this is the algorism for am and bm so that their g.c.d. is $r_n m = m(a, b)$ as the theorem requires.

A useful observation is the following:

THEOREM 3–5. Let $d = (a, b)$ be the greatest common divisor of two numbers a and b so that

$$a = a_1 d, \qquad b = b_1 d \qquad (3\text{--}3)$$

Then the two numbers a_1 and b_1 are relatively prime.

Proof: It follows from the rule in (3–2) that

$$d = (a, b) = d(a_1, b_1)$$

or $(a_1, b_1) = 1$.

This result applies in elementary arithmetic in the reduction of fractions. Any fraction

$$\frac{a}{b} = \frac{a_1}{b_1}$$

can be represented in *reduced form* with numerator and denominator that are relatively prime.

3–3. Least common multiple. A number m is said to be a *common multiple* of the numbers a and b when it is divisible by both of them. The product ab is a common multiple. Since there is only question of divisibility properties, there is no limitation in considering only the positive multiples. Among the common multiples of a and b there is a smallest one, which we shall denote by $[a, b]$ and call the *least common multiple* (l.c.m.) of a and b. The l.c.m. and the greatest common divisor have properties that in many ways are quite analogous. Corresponding to theorem 3–1 one has:

THEOREM 3–6.   Any common multiple of a and b is divisible by the least common multiple.

Proof: Let m be a common multiple of a and b. We divide m by $[a, b]$

$$m = q[a, b] + r, \qquad 0 \le r < [a, b]$$

Since m and $[a, b]$ are both divisible by a and b, it follows that the remainder r has the same property. Since $[a, b]$ is the smallest common multiple, this implies that $r = 0$ and $[a, b]$ divides m.

To determine $[a, b]$ we write a and b in the form of (3–3) where $d = (a, b)$. Any multiple of a has the form $ha = ha_1 d$. If this number is to be divisible by $b = b_1 d$, the factor ha must be divisible by b_1. Because a_1 and b_1 are relatively prime, this is possible only when h is divisible by b_1 so that $h = kb_1$. Thus any common multiple of a and b has the form

$$m = kb_1 a = ka_1 b_1 d = ka_1 b = k \frac{ab}{d}$$

For $k = 1$ one obtains the l.c.m. so that one has the result:

THEOREM 3–7.   When a and b are two numbers with the greatest common divisor $d = (a, b)$, the least common multiple is

$$[a, b] = \frac{ab}{d} \tag{3–4}$$

The formula (3–4) can be written symmetrically in regard to the l.c.m. and the g.c.d.

$$[a, b](a, b) = ab$$

Example.

Find the l.c.m. of the two numbers 76,084 and 63,020. We have already found their g.c.d. to be 92 so that

$$[76{,}084,\ 63{,}020] = \frac{76{,}084 \cdot 63{,}020}{92} = 52{,}117{,}540$$

An immediate consequence of theorem 3–7 is:

THEOREM 3-8. The least common multiple of two numbers a and b is equal to their product ab if, and only if, they are relatively prime.

Corresponding to theorem 3-4 for the g.c.d., one has the analogous formula for the l.c.m.

THEOREM 3-9.

$$[ma, mb] = m[a, b] \tag{3-5}$$

Proof: According to (3-4) and (3-2) one finds

$$[ma, mb] = \frac{ma \cdot mb}{(ma, mb)} = m\,\frac{a \cdot b}{(a, b)} = m[a, b]$$

Problem.

Determine the l.c.m. for the pairs of numbers for which the g.c.d. was found in Sec. 3-1.

3-4. Greatest common divisor and least common multiple for several numbers. So far the greatest common divisor and the least common multiple have been defined only for two numbers, but there is no difficulty in extending these concepts. Let us consider first the case of three numbers a, b, and c. A *common divisor* is any number dividing them all. Among these common divisors there is a greatest common divisor, which shall be denoted by

$$d = (a, b, c)$$

To calculate d, we observe that it is the largest number dividing c and (a, b) simultaneously, so that

$$d = ((a, b), c) \tag{3-6}$$

Example.

Let us determine the g.c.d. of the three numbers 76,084, 63,020, and 196. In a previous example we have already found $(76,084, 63,020) = 92$; consequently $d = (92, 196) = 4$.

The formula (3-6) reduces the computation of the g.c.d. of three numbers to that of two numbers. Instead of beginning with (a, b) one could have taken (b, c) first, so that

$$d = ((a, b), c) = (a, (b, c)) \tag{3-7}$$

This rule is called *the associative law* for the g.c.d.   As in theorem 3-1, one sees that every common divisor of a, b, and c divides (a, b, c).   From theorem 3-4 one concludes by means of (3-6) that

$$(ma, mb, mc) = m(a, b, c) \tag{3-8}$$

Corresponding to theorem 3-5, it follows that if one writes

$$a = a_1 d, \qquad b = b_1 d, \qquad c = c_1 d$$

then

$$(a_1, b_1, c_1) = 1$$

The g.c.d.

$$d_n = (a_1, a_2, \cdots, a_n)$$

of an arbitrary set of numbers is defined analogously.   Since d_n is the g.c.d. of a_n and the numbers $a_1, \cdots, a_{n-1}$, one concludes that

$$d_n = (d_{n-1}, a_n), \qquad d_{n-1} = (a_1, \cdots, a_{n-1})$$

This leads to a stepwise calculation

$$d_2 = (a_1, a_2), \qquad d_3 = (d_2, a_3), \cdots$$

All the rules just mentioned for the g.c.d. of three numbers hold in the general case.

Let us mention briefly the corresponding concepts for the l.c.m. A *common multiple* of three numbers a, b, and c is a number divisible by all of them.   Among these multiples there is a *least common multiple*

$$m = [a, b, c]$$

Since the l.c.m. must be divisible by $[a, b]$ and also by c, one concludes that

$$m = [[a, b], c]$$

Example.

To find the l.c.m. of the three numbers 24, 18, and 52, one calculates $[24, 18] = 72$ and $m = [72, 52] = 936$.

The l.c.m. divides all other multiples.   It obeys the *associative law*,

$$[[a, b], c] = [a, [b, c]]$$

and from theorem 3–9 one derives the rule

$$[ma, mb, mc] = m[a, b, c]$$

It is not difficult to see that when one writes

$$m = a'a = b'b = c'c$$

one must have

$$(a', b', c') = 1$$

To define and calculate the l.c.m.

$$m_n = [a_1, a_2, \cdots, a_n]$$

of a set of numbers, one can proceed stepwise as for the g.c.d.

$$m_2 = [a_1, a_2], \qquad m_3 = [m_2, a_3], \cdots$$

All properties mentioned for three numbers readily extend to this general case.

It may be recalled finally that the determination of the l.c.m. occurs naturally in elementary arithmetic in bringing fractions to their least common denominator to perform addition and subtraction.

Problems.

1. Find the g.c.d. and l.c.m. of the numbers

    (a) 63, 24, 99   (b) 16, 24, 62, 120

2. Find the l.c.m. of the integers from 1 to 10.

CHAPTER 4

PRIME NUMBERS

4–1. Prime numbers and the prime factorization theorem. An integer $p > 1$ is called a *prime number* or simply a *prime* when its only divisors are the trivial ones, ± 1 and $\pm p$. The primes below 100 are

2	13	31	53	73
3	17	37	59	79
5	19	41	61	83
7	23	43	67	89
11	29	47	71	97

The number 2 is the only *even prime*. A number $m > 1$ that is not a prime is called *composite*. The lowest composite numbers are

4	10	16
6	12	18
8	14	20
9	15	

Analogously one introduces the negative prime numbers -2, -3, -5, ..., and the negative composite numbers -4, -6, ... In the following sections we shall, as usual, consider only the positive factors in our study of the divisibility of numbers.

In regard to divisibility the primes have simple properties. We mention first:

LEMMA 4–1. A prime p is either relatively prime to a number n or divides it.

Proof: This may be concluded from the fact that the greatest common divisor of p and n is either 1 or p.

LEMMA 4-2. A product is divisible by a prime p only when p divides one of the factors.

Proof: When ab is divisible by p and a is not divisible by this prime, p is relatively prime to a and according to the division lemma must divide b. The same argument can be extended to a product of several factors.

LEMMA 4-3. A product $q_1 \ldots q_r$ of prime factors q_i is divisible by a prime p only when p is equal to one of the q_i's.

Proof: We have just seen in lemma 4-2 that p must divide some prime q_i, and since $p > 1$ one must have $p = q_i$.

LEMMA 4-4. Every number $n > 1$ is divisible by some prime.

Proof: When n is a prime, this is evident. When n is composite, it can be factored $n = ab$ where $a > 1$. The smallest possible one of these divisors a must be a prime.

We are now ready to prove the main theorem about factorizations.

THEOREM 4-1. Every composite number can be factored uniquely into prime factors.

Proof: The first step is to show that every composite number n is the product of prime factors. According to lemma 4-4 there exists a prime p_1 such that $n = p_1 n_1$. If n_1 is composite, one can draw out a further prime factor $n_1 = p_2 n_2$, and this process can be continued with the decreasing numbers n_1, n_2, $\ldots$ until some n_k becomes a prime.

After the existence of a prime factorization thus has been established, the second step consists in proving that it can only be done in one way. Let us suppose that there exist two different prime factorizations

$$n = p_1 p_2 \cdots p_k = q_1 q_2 \cdots q_l \qquad (4\text{-}1)$$

Since each p_i divides the product of the q's, it follows from lemma 4-3 that p_i is equal to some q_j, and conversely that each q is equal to some p. This shows that both sides of (4-1) contain the same primes. The only difference might be that a prime p could occur a greater number of times on one side than on the other. However, by canceling p a sufficient number of times one would obtain an equation with p on one side but not on the other, and this contradicts lemma 4–3.

The idea of the prime-factorization theorem, as well as the lemmas used in proving it, can be found in Euclid's *Elements* in Books VII and IX.

4–2. Determination of prime factors. The actual determination of the factorization of a number into prime factors is a problem of great importance in number theory. Unfortunately, for large numbers it often involves overwhelming computations.

The procedure nearest at hand consists in trying out all the lowest primes as possible divisors of the given number n. When a prime factor p has been found, one can write $n = pm$ and determine the factorization of the smaller number m. The work is limited by the previous remark that if a number is composite it must have a factor not exceeding $\sqrt{n}$, so that only primes $p \leqq \sqrt{n}$ need be divided into n. Another useful observation is that when the smallest prime factor p of n is found to be greater than $\sqrt[3]{n}$, the other factor m in $n = pm$ must be a prime. Thus if $m = ab$ were composite, both a and b would exceed $\sqrt[3]{n}$, and one would obtain the contradiction

$$n = pab > \sqrt[3]{n} \; \sqrt[3]{n} \; \sqrt[3]{n} = n$$

Example.

1. Find the prime factorization of $n = 893$. Since $\sqrt{n} < 30$ only the primes below 30 need be examined. One finds $893 = 19 \cdot 47$.

2. Find the prime factorization of the number $n = 999{,}999$. One finds successively

$$n = 3^2 \cdot 111{,}111 = 3^3 \cdot 37{,}037 = 3^3 \cdot 7 \cdot 5{,}291$$
$$= 3^3 \cdot 7 \cdot 11 \cdot 481 = 3^3 \cdot 7 \cdot 11 \cdot 13 \cdot 37$$

3. Find the prime factorization of $n = 377{,}161$. There are no obvious factors, and since $\sqrt{n} < 614$ a considerable number of primes may have to be divided into n. One finds that the smallest prime factor is $p = 137$ and $n = 137 \cdot 2{,}753$. Here the second factor is a prime since $\sqrt[3]{n} < 73$.

This method of trial and error is quite satisfactory for relatively small numbers, perhaps not exceeding four digits; for larger numbers the work involved is prohibitive, as one soon realizes. A great number of methods and devices have been invented to facilitate the determination of a factor. There exist criteria that under special circumstances make it possible to decide rather easily whether a number is a prime or not. Some of these will be mentioned later on.

Problems.

1. Find the prime factorization of the numbers: (a) 365, (b) 2,468, (c) 262,144.

2. Find the prime factorization of the two numbers: (a) 99,999, (b) 100,001.

3. Mersenne determined the factorization of the number 51,001,180,160. Find the prime factors of this number.

4–3. Factor tables. The simplest way to obtain the factorization of a number that is not too large is through the use of a *factor table*. There exist various types of these tables. The most detailed ones contain the complete factorization of every number up to some limit, but such tables are unwieldy and can give space only for relatively few numbers. To increase the capacity, most factor tables indicate only the least prime dividing each entry. Since it is quite simple to determine whether a number is divisible by the lowest primes 2, 3, 5, and 7, the numbers divisible by them are often excluded from the tables.

One of the first fair-sized factor tables was published by Rahn or Rohnius (Zurich, 1659) in an appendix to a book on algebra; the table contained the numbers up to 24,000 excluding those divisible by 2 and 5. In a translation of this work by Brancker (London, 1668) the table was extended to 100,000 by John Pell (1610–1685), an English mathematician particularly interested in number theory. For a considerable period these tables were the only ones available and they were reprinted several times in other works. The great interest in number theory in the eighteenth century created a demand for factor tables to higher limits. The strong appeals from the German scientist J. H. Lambert (1728–1777) made him a center of correspondence regarding factor tables, and several calculations were initiated. Only one of these tables was published, and even this one had an inglorious fate. It was computed by Felkel, a schoolmaster in Vienna. The first volume, which appeared in 1776 and extended to 408,000, was planned to be a part of a more ambitious program reaching several millions, most of it ready in manuscript. The tables were published at the expense of the Austrian imperial treasury, but since there was a disappointing number of subscribers, the treasury confiscated the whole edition except a couple of copies, and the paper was used in cartridges in a war against the Turks.

In the nineteenth century several large factor tables were computed by Chernac, Burckhardt, Crelle, Glaisher, and the German lightning calculator, Dase. By their combined efforts all numbers up to 10,000,000 were covered, published in individual volumes for each million. The most remarkable effort in this field was, however, the table calculated by J. P. Kulik (1773–1863), a professor of mathematics at the University of Prague. It represents the results of a twenty-year hobby and gives the factorization of the numbers up to 100,000,000. The manuscript was deposited in the library of the Vienna Academy and has not been published. The best factor table now available is the one-volume table extending to 10,000,000 prepared by D. N. Lehmer. There exist, furthermore, various special tables and punch-card devices due to D. N. Lehmer and D. H. Lehmer that greatly facilitate the determination of factors of numbers beyond the reach of tables.

4–4. Fermat's factorization method. We shall present a couple of simple methods that are sometimes very helpful in finding the factorization of a given number. The first method is due to the French mathematician and lawyer Pierre de Fermat (1601–1665), whose name we shall encounter repeatedly in the following.

Fermat must be awarded the honor of being the founding father of number theory as a systematic science. His life was quiet and uneventful and entirely centered around the town of Toulouse, where he first studied jurisprudence, practiced law, and later became prominent as councilor of the local parliament. His leisure time was devoted to scholarly pursuits and to a voluminous correspondence with contemporary mathematicians, many of whom, like himself, were gentlemen-scholars, the ferment of intellectual life in the seventeenth and eighteenth centuries. Fermat possessed a broad knowledge of the classics, enjoyed literary studies, and wrote verse, but mathematics was his real love. He published practically nothing personally, so that his works have been gleaned from notes that were preserved after his death by his family, and from letters and treatises that he sent to his correspondents. In spite of his modesty, Fermat gained an outstanding reputation for his mathematical achievements. He made considerable contributions to the foundation of the theory of probability in his correspondence with Pascal and introduced coordinates independent of Descartes. The French, when too exasperated over the eternal priority squabble between the followers of Newton and Leibniz, often interject the name of Fermat as a cofounder of the calculus. There is considerable justification for this point of view. Fermat did not reduce his procedures to rule-of-thumb methods, but he did perform a great number of differentiations by tangent determinations and integrations by computations of numerous areas, and he actually gave methods for finding maxima and minima corresponding to those at present used in the differential calculus.

In spite of all these achievements, Fermat's real passion in mathematics was undoubtedly number theory. He returned to such problems in almost all his missives; he delighted to propose new and difficult problems, and to give solutions in large figures that require elaborate computations; and most important of all, he announced new principles and methods that have inspired all work in number theory after him.

Fermat's factorization method, which is the point interesting us particularly for the moment, is found in an undated letter of about 1643, probably addressed to Mersenne (1588–1648). Mersenne was a Franciscan friar and spent most of his lifetime in cloisters in Paris. He was an aggressive theologian and philosopher, a schoolmate and close friend of Descartes. He wrote some mathematical works, but a greater part of his importance in the history of mathematics rests on the fact that he was a favorite intermediary in the correspondence between the most prominent mathematicians of the times.

Fermat's method is based upon the following facts. If a number n can be written as the difference between two square numbers, one has the obvious factorization

$$n = x^2 - y^2 = (x - y)(x + y) \qquad (4\text{-}2)$$

On the other hand when

$$n = ab, \ b \geqq a$$

is composite, one can obtain a representation (4-2) of n as the difference of two squares by putting

$$x - y = a, \qquad x + y = b$$

so that

$$x = \frac{b + a}{2}, \qquad y = \frac{b - a}{2} \qquad (4\text{-}3)$$

Since we deal with the question of factoring n, we can assume that n is odd; hence a and b are odd and the values of x and y are integral.

Corresponding to each factorization of n there exists, therefore, a representation (4-2). To determine the possible x and y in (4-2), we write

$$x^2 = n + y^2$$

Since $x^2 \geqq n$ one has $x \geqq \sqrt{n}$. The procedure consists in substituting successively for x the values above $\sqrt{n}$ and examining whether the corresponding

$$\Delta(x) = x^2 - n$$

is a square y^2. Let us illustrate by a simple example.

Example.

The number $n = 13{,}837$ is to be factored. One sees that $\sqrt{n}$ lies between 117 and 118. In the first step we obtain

$$\Delta(118) = 118^2 - 13{,}837 = 87$$

which is not a square. In the next step one has

$$\Delta(119) = 119^2 - 13{,}837 = 324 = 18^2$$

so that we have found the factorization

$$13{,}837 = (119 - 18)(119 + 18) = 101 \cdot 137$$

This example is too simple to illustrate the short cuts that serve to facilitate the work with larger numbers. One important observation is that one need not calculate each $\Delta(x)$ separately. Since

$$(x + 1)^2 - n = x^2 - n + 2x + 1$$

one has

$$\Delta(x + 1) = \Delta(x) + 2x + 1$$

and by applying this rule repeatedly one finds

$$\Delta(x + 2) = \Delta(x + 1) + 2x + 3$$

$$\Delta(x + 3) = \Delta(x + 2) + 2x + 5$$

. . . .

This makes it possible to compute the successive $\Delta(x)$'s by simple additions.

Example.

We shall take the formidable number $n = 2{,}027{,}651{,}281$, on which Fermat applied his method. The first integer above $\sqrt{n}$ is 45,030 and the calculations proceed as follows:

$x = 45{,}030$	$x^2 - n =$	49,619
	$2x + 1 =$	90,061
31		139,680
		90,063
32		229,743
		90,065
33		319,808
		90,067
34		409,875
		90,069
45,035		499,944

$$
\begin{array}{rl}
45,035 & 499,944 \\
 & 90,071 \\ \hline
36 & 590,015 \\
 & 90,073 \\ \hline
37 & 680,088 \\
 & 90,075 \\ \hline
38 & 770,163 \\
 & 90,077 \\ \hline
39 & 860,240 \\
 & 90,079 \\ \hline
45,040 & 950,319 \\
 & 90,081 \\ \hline
x = 45,041 & 1,040,400 = 1,020^2 = y^2
\end{array}
$$

This shows that we have the factorization

$$n = (45,041 + 1,020)(45,041 - 1,020)$$

$$= 46,061 \cdot 44,021$$

where each factor can be shown to be a prime. In this chain of computations, each of the various numbers 49,619, 139,680, ... should be looked up in a table of squares to determine whether it is actually a perfect square. However, in most cases this step may be eliminated since the last two digits will already show that the number is not a square. The small table of 22 entries that we computed on page 33 giving the possible two last digits of a square number is most convenient for this purpose. Of all the numbers in the preceding chain it is only necessary to look up the numbers 499,944 and 1,040,400, since 44 and 00 may be the last two digits in a square.

Fermat's method is particularly helpful when the number n has two factors whose difference

$$2y = b - a$$

is relatively small, because a suitable y will then quickly appear. In the choice of the example discussed above it is clear that Fermat had this in mind. By means of certain other improvements that can be introduced in the procedure, it becomes one of the most effective factorization methods available.

Problem.

Factor the following numbers by means of Fermat's method: (a) 8,927, (b) 57,479, (c) 14,327,581.

4–5. Euler's factorization method. Frénicle de Bessy (1605–1675) was an official at the French mint and was well known for his unusual facility in numerical computations. He was also a mathematician of no mean ability and was in frequent corre-

Fig. 4–1. Leonhard Euler (1707–1783).

spondence with Fermat. In a letter of August 2, 1641, he proposes the following problem: Use the fact that

$$221 = 10^2 + 11^2 = 5^2 + 14^2 \qquad (4\text{–}4)$$

to find the factors of this number. The same idea, that two different representations of a number as a sum of two squares may serve to factor it, was mentioned by Mersenne. However, Euler, for whom the method is usually named, seems to have been the first to put it to extensive use.

Leonhard Euler (1707–1783) was a remarkable scientist whose contributions have left their imprint on almost all branches of mathematics. His papers were rewarded ten times by prizes of the French Academy. His productivity was immense; it has been estimated that his collected works, which are still in the process of being published, will fill upward of 100 large volumes. Euler was born in Switzerland, but he was early called to the Academy in St. Petersburg, later to the Academy in Berlin at the request of Frederic II, and back again to St. Petersburg on still more flattering terms.   As a young man he lost the sight of one eye and later in life he became totally blind, but even this calamity did not halt his scientific work.   One of his best known texts, *Complete Introduction to Algebra* (1770), which contains much material on elementary number theory, was dictated to a servant, a former tailor, to prepare him to serve as his mathematical secretary.

Euler carried on an extensive correspondence with contemporary mathematicians, and the factorization by means of representation of a number as the sum of two squares is mentioned first in a letter of February 16, 1745, to Christian Goldbach (1690–1764).   Goldbach was a German mathematician, onetime teacher of Peter II and secretary for the Academy in St. Petersburg, who left scientific work to embark upon a distinguished career in the Russian civil service.

Euler's factorization method applies only to numbers which in some way can be represented as a sum of two squares

$$N = a^2 + b^2 \qquad (4\text{–}5)$$

as, for instance,

$$41 = 5^2 + 4^2, \qquad 269 = 10^2 + 13^2$$

Since we may assume that the number N to be factored is odd, one of the numbers in (4–5), say a, is odd and the other, b, is even. We have observed that the square of an odd number a^2 is of the form $4n + 1$, and since b^2 is divisible by 4, the number N itself must be of the form $4m + 1$.

We shall assume now that there exists another representation of N as the sum of two squares

$$N = c^2 + d^2 \qquad (4\text{–}6)$$

as, for instance, in the example (4–4) given by Frénicle.   The notation is again such that c is odd and d even.   To show that the two representations lead to a factorization of N, we proceed as follows. From (4–5) and (4–6) we have

$$a^2 + b^2 = c^2 + d^2$$

so that

$$a^2 - c^2 = d^2 - b^2$$

or

$$(a - c)(a + c) = (d - b)(d + b) \qquad (4\text{–}7)$$

Let k be the greatest common factor of $a - c$ and $d - b$ so that

$$a - c = kl, \qquad d - b = km, \qquad (l, m) = 1 \qquad (4\text{–}8)$$

By our choice of notations $a - c$ and $d - b$ are even, hence k is even.   When (4–8) is substituted into (4–7) and k is canceled, one obtains

$$l(a + c) = m(d + b) \qquad (4\text{–}9)$$

Since l and m are relatively prime, $a + c$ must be divisible by m

$$a + c = mn \qquad (4\text{–}10)$$

When this is applied in (4–9), finally

$$d + b = ln \qquad (4\text{–}11)$$

The two expressions (4–10) and (4–11) also show that n is the g.c.d. of $a + c$ and $d + b$; thus n is even.

The desired factorization of N which results from (4–5) and (4–6) is now

$$N = \left[\left(\frac{k}{2}\right)^2 + \left(\frac{n}{2}\right)^2\right](m^2 + l^2) \qquad (4\text{–}12)$$

To prove that this equation is correct, we multiply out the expression on the right-hand side and find that it is equal to

$$\tfrac{1}{4}[(km)^2 + (kl)^2 + (nm)^2 + (nl)^2]$$

Here we substitute the values from (4–8), (4–10), and (4–11) so that the new expression becomes

$$\tfrac{1}{4}[(d-b)^2 + (a-c)^2 + (a+c)^2 + (d+b)^2]$$
$$= \tfrac{1}{4}(2a^2 + 2b^2 + 2c^2 + 2d^2) = \tfrac{1}{4}(2N+2N) = N$$

as we required.

Example.

For the number $N = 221$ the two representations (4–4) yield

$$a = 11, \qquad a - c = \ 6, \qquad k = 2$$
$$b = 10, \qquad a + c = 16, \qquad l = 3$$
$$c = \ 5, \qquad d - b = \ 4, \qquad m = 2$$
$$d = 14, \qquad d + b = 24, \qquad n = 8$$

The decomposition (4–12) is therefore

$$221 = (1 + 4^2)(2^2 + 3^2) = 17 \cdot 13$$

Clearly the decomposition (4–12) is never trivial in the sense that any of the factors is equal to 1. To apply Euler's method one has to determine two representations of a number as a sum of two squares. This may be done by means of tables of squares, as in the case of Fermat's method. Often the number is given in such a form that one representation is immediate. To find any representation of a number N as the sum of two squares, one forms the differences $N - x^2$ for various x's and examines whether they can be squares y^2. Many cases are immediately excluded by inspection of the last two digits of $N - x^2$. As before, one can reduce the calculations to additions by observing that one obtains $N - (x-1)^2$ from $N - x^2$ simply by adding $2x - 1$.

Examples.

1. Let us factor $N = 2{,}501$. Since $N = 50^2 + 1$ we need only another such representation. One finds for $x = 50$,

$$2{,}501 - x^2 = \ \ 1$$
$$\underline{\ \ \ 2x - 1 \ = \ 99 \ \ }$$
$$2{,}501 - 49^2 = 100 = 10^2$$

Thus one has

$$a = 1, \quad a - c = -48, \quad k = 8$$
$$b = 50, \quad a + c = 50, \quad l = -6$$
$$c = 49, \quad d - b = -40, \quad m = -5$$
$$d = 10, \quad d + b = 60, \quad n = 10$$

so that the decomposition (4–12) is

$$2,501 = (4^2 + 5^2)(5^2 + 6^2) = 41 \cdot 61$$

2. Euler applied his method to decompose

$$N = 1,000,009 = 1,000^2 + 3^2$$

He finds a second representation

$$N = 972^2 + 235^2$$

and this leads to the factorization

$$N = 293 \cdot 3,413$$

It is possible to show that if a number can be represented as the sum of two squares one can find all factorizations by Euler's method. Euler succeeded also in obtaining a proof for the following theorem due to Fermat: Every prime of the form $4n + 1$ can be represented as the sum of two squares. From our preceding results we conclude that such a representation can be made in only one way, since otherwise the number would be factorable. The proof of the theorem of Fermat will be given in Chap. 11. Let us illustrate the theorem on the primes of the form $4n + 1$ below 100:

$$5 = 2^2 + 1^2, \quad 13 = 3^2 + 2^2, \quad 17 = 4^2 + 1^2, \quad 29 = 5^2 + 2^2$$
$$37 = 6^2 + 1^2, \quad 41 = 5^2 + 4^2, \quad 53 = 7^2 + 2^2, \quad 61 = 6^2 + 5^2$$
$$73 = 8^2 + 3^2, \quad 89 = 8^2 + 5^2, \quad 97 = 9^2 + 4^2$$

Euler's factorization method is capable of wide extensions. It leads to the theory of representations of numbers by means of quadratic forms, *i.e.*,

$$N = ax^2 + bxy + cy^2$$

Such representations can under certain conditions be used for factoring in the same manner as the special form

$$N = x^2 + y^2$$

It would carry us too far to discuss the great number of other aids and methods for factoring, some of them very ingenious. We shall make only a final remark about the last digits of factors. If, for instance, N has the last digit 1, one finds by checking all possibilities that the two eventual factors must both end in 1, or both in 9, or one in 3 and the other in 7. When other last digits in N are examined, one finds the following table:

Last Digit in Number	Last Digit in Factors
1	(1, 1), (9, 9), (3, 7)
3	(1, 3), (7, 9)
7	(1, 7), (3, 9)
9	(1, 9), (3, 3), (7, 7)

The remaining digits 0, 2, 4, 5, 6, 8 are of no interest since there is an obvious factor 2 or 5 in N. This method may be extended in various ways, for instance, to several digits or to representations of the number in number systems with a basis different from 10.

Problems.

1. Factor the numbers (*a*) 19,109, (*b*) 10,001 by Euler's method.

2. Express all primes of the form $4n + 1$ between 100 and 200 as the sum of two squares.

4–6. The sieve of Eratosthenes. The factorization theorem states that every number can be represented uniquely as the product of prime factors. Thus the prime numbers, as their name already indicated in the Greek terminology, are the first building stones from which all other numbers may be created multiplicatively. As a consequence considerable efforts have been concentrated on the study of primes.

The first result that we shall mention has been derived in Euclid's *Elements* (Proposition 20, Book IX).

THEOREM 4–2. There is an infinitude of primes.

Euclid's proof runs as follows: Let a, b, c, . . . , k be any family of prime numbers. Take their product $P = ab \cdots k$ and add 1. Then $P + 1$ is either a prime or not a prime. If it is, we have added another prime to those given. If it is not, it must be divisible by some prime p. But p cannot be identical with any of the given prime numbers a, b, . . . , k because then it would divide P and also $P + 1$; hence it would divide their difference, which is 1, and this is impossible. Therefore a new prime can always be found to any given (finite) set of primes.

We may illustrate the construction of primes by Euclid's method by the following examples:

$2 \cdot 3 + 1 = 7 = \text{prime}$

$2 \cdot 3 \cdot 5 + 1 = 31 = \text{prime}$

$2 \cdot 3 \cdot 5 \cdot 7 + 1 = 211 = \text{prime}$

$2 \cdot 3 \cdot 5 \cdot 7 \cdot 11 + 1 = 2{,}311 = \text{prime}$

$2 \cdot 3 \cdot 5 \cdot 7 \cdot 11 \cdot 13 + 1 = 30{,}031 = 59 \cdot 509$

$2 \cdot 3 \cdot 5 \cdot 7 \cdot 11 \cdot 13 \cdot 17 + 1 = 510{,}511 = 19 \cdot 97 \cdot 277$

$2 \cdot 3 \cdot 5 \cdot 7 \cdot 11 \cdot 13 \cdot 17 \cdot 19 + 1 = 9{,}699{,}691 = 347 \cdot 27{,}953$

In Euclid's proof one could just as well have used the number $P - 1$. When applied to the first primes, this leads to the factorizations

$2 \cdot 3 - 1 = 5 = \text{prime}$	$30{,}029 = \text{prime}$
$2 \cdot 3 \cdot 5 - 1 = 29 = \text{prime}$	$510{,}509 = 61 \cdot 8{,}369$
$209 = 11 \cdot 19$	$9{,}699{,}689 = 53 \cdot 197 \cdot 929$
$2{,}309 = \text{prime}$	

There are many other numbers which could have served in a

similar manner to obtain arbitrarily large prime factors, for instance, $n! \pm 1$, where as usual

$$n! = 1 \cdot 2 \cdot 3 \cdots n$$

is n *factorial*. The reader may try to factor some of these numbers.

Extensive tables of primes have been computed. Clearly every factor table gives information about the primes within its range, but it is desirable also to have separate lists of primes. Generally available and unusually free from errors are the tables of primes up to 10,000,000 prepared by D. N. Lehmer (1867–1938).

There exists an ancient method of finding the primes known as the *sieve of Eratosthenes*. Eratosthenes (276–194 B.C.) was a Greek scholar, chief librarian of the famous library in Alexandria. He is noted for his chronology of ancient history and for his measurement of the meridian between Assuan and Alexandria, which made it possible to estimate the dimensions of the earth with fairly great accuracy.

Eratosthenes' sieve method consists in writing down all numbers up to some limit, say 100:

1	2	3	4̄	5	6̿	7	8̄	9̄	10̄	11	12̿	13	14̿	15̿	16̄
	17	18̄	19	20̿	21̄	22̄	23	24̿	25̄	26̄	27̄	28̿	29	30̿	31
	32̄	33̄	34̄	35̿	36̿	37	38̄	39̄	40̄	41	42̿	43	44̄	45̿	46̄
	47	48̿	49̄	50̿	51̄	52̄	53	54̿	55̄	56̄	57̄	58̄	59	60̿	61
	62̄	63̿	64̄	65̄	66̿	67	68̄	69̄	70̿	71	72̿	73	74̄	75̿	76̄
	77̄	78̿	79	80̿	81̄	82̄	83	84̿	85̄	86̄	87̄	88̄	89	90̿	91̄
	92̄	93̄	94̄	95̄	96̿	97	98̄	99̄	100̄						

From this series one first strikes out every second number counting from 2, that is, the numbers 4, 6, 8, ... In the example above these numbers are marked by a bar. Counting from the first remaining number, 3, every third number, that is, 6, 9, 12, . . . is marked; some of them will thus have a second bar. The next remaining number is 5, which is a prime since it has not been struck out as divisible by 2 or 3; every fifth number 10, 15, 20, . . . is eliminated. The first remaining number, 7, is a prime since it is not divisible by 2, 3, or 5, and its multiples 14, 21, . . . are marked. In this manner all primes may be determined successively. Clearly the method is well adapted to mechanical procedures. Since it is not necessary to write the numbers explicitly, one can use stencils or punch cards. All larger factor and prime tables have been constructed by means of such devices.

The following observation is essential in the application of Eratosthenes' sieve. In the preceding example, when all multiples of 7 have been marked in the fourth step, the remaining unmarked numbers will now include all primes below 100, since no remaining number N has any factor less than the next prime $11 > \sqrt{N}$.

This fact makes it possible to use the sieve of Eratosthenes to calculate the number of primes up to prescribed limits. It is customary to denote the number of primes not exceeding a number x by $\pi(x)$; for instance, $\pi(12) = 5$, $\pi(17) = 7$. Let us return to our example again. Here we had four primes below $\sqrt{100}$, namely, 2, 3, 5, and 7. Let us perform the canceling in a slightly different manner so that in the first step also the prime 2 is eliminated, in the second step also 3 is canceled, and so on. What is left after the four cancellations of multiples of 2, 3, 5, and 7 will be the number 1 and the primes between 10 and 100, hence altogether

$$\pi(100) - \pi(\sqrt{100}) + 1$$

numbers. On the other hand in the first step one cancels $100/2$ numbers out of 100. In the second step one cancels $[100/3]$, recalling that the bracket denotes the integral part of the quotient. There is, however, some duplication since the $[100/(2 \cdot 3)]$ numbers divisible by both 2 and 3 have been eliminated twice. After the two steps there remain consequently

$$100 - \left[\frac{100}{2}\right] - \left[\frac{100}{3}\right] + \left[\frac{100}{2 \cdot 3}\right]$$

numbers. In the third application of the sieve one eliminates

[100/5] numbers, but there is duplication with respect to those divisible by $2 \cdot 5$ and $3 \cdot 5$ so that the next further reduction is

$$\left[\frac{100}{5}\right] - \left[\frac{100}{2 \cdot 5}\right] - \left[\frac{100}{3 \cdot 5}\right] + \left[\frac{100}{2 \cdot 3 \cdot 5}\right]$$

where the last term takes care of the fact that those numbers that are divisible by $2 \cdot 3 \cdot 5$ have been subtracted twice from [100/5]. Thus we conclude that out of the 100 original numbers there is now left a total of

$$100 - \left[\frac{100}{2}\right] - \left[\frac{100}{3}\right] - \left[\frac{100}{5}\right] + \left[\frac{100}{2 \cdot 3}\right]$$
$$+ \left[\frac{100}{2 \cdot 5}\right] + \left[\frac{100}{3 \cdot 5}\right] - \left[\frac{100}{2 \cdot 3 \cdot 5}\right]$$

After the fourth step one verifies similarly that there remains

$$\tau(100) - \pi(10) + 1 = 100 - \left[\frac{100}{2}\right] - \left[\frac{100}{3}\right] - \left[\frac{100}{5}\right] - \left[\frac{100}{7}\right]$$
$$+ \left[\frac{100}{2 \cdot 3}\right] + \left[\frac{100}{2 \cdot 5}\right] + \left[\frac{100}{2 \cdot 7}\right] + \left[\frac{100}{3 \cdot 5}\right] + \left[\frac{100}{3 \cdot 7}\right] + \left[\frac{100}{5 \cdot 7}\right]$$
$$- \left[\frac{100}{2 \cdot 3 \cdot 5}\right] - \left[\frac{100}{2 \cdot 3 \cdot 7}\right] - \left[\frac{100}{2 \cdot 5 \cdot 7}\right] - \left[\frac{100}{3 \cdot 5 \cdot 7}\right] + \left[\frac{100}{2 \cdot 3 \cdot 5 \cdot 7}\right]$$

This makes it possible to calculate $\pi(100)$, since $\pi(10) = 4$ so that

$$\pi(100) - 3 = 100 - (50 + 33 + 20 + 14)$$
$$+ (16 + 10 + 7 + 6 + 4 + 2) - (3 + 2 + 1 + 0) + 0 = 22$$

or $\pi(100) = 25$ as one could have counted directly from the table of primes.

It is clear that through the preceding considerations we have been led to a general formula regarding the number of primes.

Let N be the given number and $p_1, p_2, \ldots, p_r$ the primes less than $\sqrt{N}$. Then

$$\pi(N) - \pi(\sqrt{N}) + 1 = N - \left[\frac{N}{p_1}\right] - \left[\frac{N}{p_2}\right] - \cdots - \left[\frac{N}{p_r}\right]$$
$$+ \left[\frac{N}{p_1 p_2}\right] + \left[\frac{N}{p_1 p_3}\right] + \cdots + \left[\frac{N}{p_{r-1} \cdot p_r}\right]$$
$$- \left[\frac{N}{p_1 p_2 p_3}\right] - \cdots + \cdots$$

It is not difficult to prove this result in general by means of induction. Through the formula one can determine the number of primes below N when the primes below $\sqrt{N}$ are known. The method is cumbersome; for instance, to find the number of primes below 10,000 one must consider the primes less than 100. Some simplification is derived from the fact that many of the terms must vanish. However, as shown by Meissel (1870), the formula may be considerably improved, and through various short cuts he succeeded in finding

$$\pi(100,000,000) = 5,761,455$$

These computations were continued by the Danish mathematician Bertelsen, who applied them for the determination of errors in prime tables. He announced the following result (1893):

$$\pi(1,000,000,000) = 50,847,478$$

which represents our most extended knowledge of the number of primes.

Problem.

Determine the number of primes below 200 by the method given above and check the result by actual count.

4–7. Mersenne and Fermat primes. Considerable effort has been centered on the factorization of numbers of particular types. Some of them are numbers resulting from mathematical problems

of interest.   Others have been selected because it is known for
theoretical reasons that the factors must have a special form.
Among the numbers that have been examined in great detail one
should mention the so-called *binomial numbers*

$$N = a^n \pm b^n \tag{4-13}$$

where a and b are integers.   Certain factors can be obtained
immediately from their algebraic expression, since

$$a^n - b^n = (a - b)(a^{n-1} + a^{n-2}b + \cdots + ab^{n-2} + b^{n-1}) \tag{4-14}$$

as one verifies by performing the right-hand multiplication.   By
putting $-b$ for b in (4–14), one obtains for odd exponents n

$$a^n + b^n = (a + b)(a^{n-1} - a^{n-2}b + \cdots - ab^{n-2} + b^{n-1}) \tag{4-15}$$

If one replaces a and b in (4–14) by a^m and b^m, it follows that

$$a^{nm} - b^{nm} = (a^m - b^m)(a^{(n-1)m} + a^{m(n-2)}b^m + \cdots + b^{m(n-1)}) \tag{4-16}$$

This expression may be used to factor a number (4–13) when the
exponent is composite.   Thus, every number (4–13) has certain
algebraic factors that are fairly easily found, and the essential
difficulty lies in factoring these further or in establishing that they
are primes.   Here one is aided by some knowledge of the type of
primes that can divide them.

Examples.

1. Factor

$$N = 10^9 - 3^9 = 999,980,317$$

One finds the algebraic factors

$$10 - 3 = 7 \quad \text{and} \quad 10^3 - 3^3 = 973 = 7 \cdot 139$$

By using a factor table on the remaining factor, one finds the prime decomposition

$$N = 7 \cdot 19 \cdot 139 \cdot 54,091$$

2. The number

$$N = 10^9 + 3^9 = 1,000,019,683$$

has the algebraic factors

$$10 + 3 = 13 \qquad \text{and} \qquad 10^3 + 3^3 = 1{,}027 = 13 \cdot 79$$

and the final result is

$$N = 13 \cdot 37 \cdot 79 \cdot 26{,}317$$

The prime factorization of the numbers

$$M_n = 2^n - 1 \tag{4-17}$$

has been the object of intensive studies. Their decomposition is known and tabulated for a large number of exponents n. For small exponents the reader can easily determine the factors; for instance,

$$M_2 = 3 \qquad\qquad M_6 = 63 = 3 \cdot 3 \cdot 7$$

$$M_3 = 7 \qquad\qquad M_7 = 127$$

$$M_4 = 15 = 3 \cdot 5 \qquad M_8 = 255 = 3 \cdot 5 \cdot 17$$

$$M_5 = 31$$

As an example of a more imposing factorization, let us give a prime decomposition that the French mathematician Poulet worked on as a pastime during the occupation in the Second World War.

$$2^{135} - 1 = 7 \cdot 31 \cdot 73 \cdot 151 \cdot 271 \cdot 631 \cdot 23{,}311$$
$$\cdot\, 262{,}657 \cdot 348{,}031 \cdot 49{,}971{,}617{,}830{,}801$$

The reason for the particular interest in the numbers (4–17) can be found in the fact that they are directly associated with the classical problem of the *perfect numbers*, which we shall discuss in the next chapter. Every number M_n that is a prime gives rise to a perfect number. These primes are known as *Mersenne primes*. The historical justification for this nomenclature seems rather weak, since several perfect numbers and their corresponding primes have been known since antiquity and occur in almost every medieval numerological speculation. Mersenne did, however, discuss the primes named after him in a couple of places in his work *Cogita physico-mathematica* (Paris, 1644) and expressed various conjectures in regard to their occurrence.

It is clear that a number of the type (4–17) cannot be a prime when $n = rs$ is composite, because there would exist an algebraic factorization of M_n as in (4–16).   Since in this case $a - b = 2 -$

FIG. 4–2.   Marin Mersenne (1588–1648).

$1 = 1$, the factorization (4–14) is trivial.   One concludes therefore that a Mersenne prime has the form

$$M_p = 2^p - 1$$

where the exponent p is itself a prime.   As a consequence, these numbers have been investigated for many primes p.   For small p one finds relatively many Mersenne primes, but for larger p they seem to become more and more scarce.   At present only 12 Mersenne primes are known;   the first ones are

$$M_2 = \quad 3 \qquad M_{13} = \quad 8{,}191$$
$$M_3 = \quad 7 \qquad M_{17} = 131{,}071$$
$$M_5 = \quad 31 \qquad M_{19} = 524{,}287$$
$$M_7 = 127$$

The last two of these were determined by the early Italian mathematician Cataldi (1552–1626) in his *Trattato de numeri perfetti* by the direct procedure of dividing by all primes less than the square root of the number. Cataldi was an enthusiastic protagonist for mathematical studies. He founded the first mathematical academy in his native town Bologna and distributed his works free in Italian cities to create interest in the subject.

The next Mersenne prime M_{31} was determined by Euler (1750); another, M_{61}, by Pervouchine (1883) and Seelhoff (1886). Powers (1911) found that M_{89} and, later (1914), M_{107} are primes. The largest and last of the known Mersenne primes is

$$M_{127} = 170{,}141{,}183{,}460{,}469{,}231{,}731{,}687{,}303{,}715{,}884{,}105{,}727$$

The only reason for writing explicitly this huge number of 39 digits is that it is the largest number that has actually been verified to be a prime. It was found by the French mathematician Lucas in 1876. Lucas (1842–1891) discovered a new and very much simpler method for testing the primality of the Mersenne numbers. They have now been examined by means of Lucas's criterion for the primes up to and including $p = 257$, and no new Mersenne primes have been found. The examination of the last few remaining ones up to this limit has just been completed by H. S. Uhler. (See Supplement.)

A family of numbers related to the Mersenne numbers are those of the form

$$N_n = 2^n + 1 \tag{4-18}$$

Fermat initiated the study of their factors and their primality. Now, for a number of the type (4–18) to be a prime, it is clear that the exponent n cannot have any odd factor. If, for example, $n = ab$ where b is odd, one would obtain an algebraic factorization as in (4–15)

$$2^n + 1 = (2^a)^b + 1$$
$$= (2^a + 1)(2^{a(b-1)} - 2^{a(b-2)} + 2^{a(b-3)} - \cdots + 1)$$

However, a number without odd factors must be a power of 2 so that $n = 2^t$, and the numbers take the form

$$F_t = 2^{2^t} + 1 \qquad (4\text{--}19)$$

These numbers are known as the *Fermat numbers*, and for the first values of t they are seen to be primes

$$F_0 = 3, \qquad F_1 = 5, \qquad F_2 = 17, \qquad F_3 = 257, \qquad F_4 = 65{,}537$$

The next Fermat number is already so large that it is difficult to factor, but on the basis of the few facts at hand Fermat made the conjecture that they are all primes. He expresses this conjecture repeatedly, in letters to Frénicle, Pascal, and others. In August, 1640, he states: "Je n'en ai pas la démonstration exacte, mais j'ai exclu si grande quantité de diviseurs par démonstrations infaillibles, et j'ai de si grandes lumières, qui établissent ma pensée que j'aurois peine à me dédire."

It was not until 100 years later (1739) that Euler exploded the hypothesis by the simple expedient of showing that the next Fermat number had a factor. Euler showed first theoretically that any factor of a Fermat number must have the form

$$2^{t+1}k + 1$$

For $t = 5$ one concludes, therefore, that the prime factors must have the form $p = 64k + 1$. From a prime table one finds that the first primes of this kind are 193, 257, 449, 577, and finally 641, which actually turns out to be a factor of F_5.

Through this discovery the Fermat numbers lost much of their attraction and actuality as a research object. However, through one of the peculiar twists of the lines of mathematical investigation, they reappeared with greater importance in an unsuspected and quite surprising connection with a classical problem. In his famous *Disquisitiones arithmeticae* the German mathematician C. F. Gauss in 1801 among other things took up the ancient problem of finding all regular polygons that can be constructed by means of compass and ruler. We shall return to the *Disquisitiones* and the problem of the regular polygons later on. It must suffice to state here that after the investigations of Gauss, the problem was reduced to the question of the existence of the Fermat primes. As a consequence, they have been the object of numerous studies, both theoretical and computational, and quite a few of the larger Fermat numbers have been successfully factored. Of Fermat's original conjecture there is no trace; no further primes have been found. Students of the question now seem more inclined to the opposite hypothesis that there are no further Fermat primes than the first five already found. A survey of the present state of the factorizations of Fermat and Mersenne numbers can be found in a recent paper by D. H. Lehmer.[1]

Problems.

1. Factor the numbers

$$10^8 \pm 3^8$$

2. Factor some of the first of the numbers

$$2^n \pm 1$$

beyond those given above.

4–8. The distribution of primes. By checking the entries in a prime table one sees soon that aside from minor irregularities the prime numbers gradually become more scarce. The sieve of Eratosthenes shows that this must be the case since in the higher intervals more and more numbers become effaced. For instance, by actual count one finds that each hundred from 1 to 1,000 contains respectively the following number of primes:

$$25, 21, 16, 16, 17, 14, 16, 14, 15, 14$$

while in the hundreds from 1,000,000 to 1,001,000, the corresponding frequencies are

$$6, 10, 8, 8, 7, 7, 10, 5, 6, 8$$

and from 10,000,000 to 10,001,000,

$$2, 6, 6, 6, 5, 4, 7, 10, 9, 6$$

[1] LEHMER, D. H., "On the Factors of $2^n \pm 1$," *Bulletin of the American Mathematical Society*, Vol. 53, 164–167 (1947).

A special computation by M. Kraitchik shows that for the interval from 10^{12} to $10^{12} + 1{,}000$ the corresponding figures are

$$4, 6, 2, 4, 2, 4, 3, 5, 1, 6$$

Except for the case $p = 2$ the primes are odd, so any two consecutive primes must have a distance that is at least equal to 2. Pairs of primes with this shortest distance are called *prime twins*; for instance,

$$(3, 5), (5, 7), (11, 13), (17, 19), (29, 31), \cdots,$$

$$(10{,}006{,}427, \ 10{,}006{,}429)$$

In spite of the fact that these prime twins become quite rare in the tables, it is still believed that their number is infinite.   On the other hand, one can also find consecutive primes whose distance is as large as one may wish; in other words, there exist arbitrarily long sequences of numbers that are all composite.   To prove this statement one need only observe that when

$$n! = 1 \cdot 2 \cdot 3 \cdot \ \cdots \ \cdot n$$

the $n - 1$ numbers

$$n! + 2, \quad n! + 3, \quad \cdots, \quad n! + n$$

are all composite.

These remarks show that there are great irregularities in the occurrence of the primes.   Nevertheless, when the large-scale distribution of primes is considered, it appears in many ways quite regular and obeys simple laws.   The study of these laws in the distribution of the primes falls in the field of analytic number theory.   This particular domain of number theory operates with very advanced methods of the calculus and is considered to be technically one of the most difficult fields of mathematics.   Its central problem is the study of the function $\pi(x)$, which indicates the number of primes up to a certain number x.   It was discovered quite early by means of empirical counts in the prime tables that the function $\pi(x)$ behaves asymptotically like the function $x/\log x$, that is, for large values of x their quotient approaches 1

$$\lim_{x \to \infty} \frac{\pi(x)}{x/\log x} = 1 \qquad (4\text{-}20)$$

(The logarithm here and in the following is the natural logarithm to the base e.)

This does not, of course, mean that the difference $\pi(x) - \dfrac{x}{\log x}$ becomes small, but only that this difference is small in comparison with $\pi(x)$. The result that is expressed in the formula (4-20) is commonly known as the *prime-number theorem*.

For the purpose of approaching the prime function $\pi(x)$, the so-called *integral logarithm* is better than the function $x/\log x$, although for large values of x the two functions behave asymptotically alike. The integral logarithm is defined by means of an integral

$$\mathrm{Li}(x) = \int_2^x \frac{dt}{\log t}$$

The following table indicates the accuracy of the approximation:

x	$\pi(x)$	$\mathrm{Li}(x)$
1,000	168	178
10,000	1,229	1,246
100,000	9,592	9,630
1,000,000	78,498	78,628
10,000,000	664,579	664,918
100,000,000	5,761,455	5,762,209
1,000,000,000	50,847,478	50,849,235

Here the values of $\pi(x)$ up to 10,000,000 have been obtained by actual count from tables of primes, while the two remaining entries are the values of $\pi(x)$ calculated by Meissel and Bertelsen, which we have mentioned previously.

Already Euler had begun applying the methods of the calculus to number-theory problems. However, the German mathematician G. F. B. Riemann (1826–1866) is generally regarded as the real founder of analytic number theory. His personal life was modest and uneventful until his premature death from tuberculosis. According to the wish of his father he was originally destined to become a minister, but his shyness and lack of ability as a speaker made him abandon this plan in favor of mathematical scholarship. He was unassuming to a fault, yet at present he is recognized as one of the most penetrating and original mathematical minds of the nineteenth century. In analytic number theory, as well as in many other fields of mathematics, his ideas still have a profound influence. His starting point was a function now called *Riemann's zeta function*

$$\zeta(s) = 1 + \frac{1}{2^s} + \frac{1}{3^s} + \frac{1}{4^s} + \cdots$$

This function he investigated in great detail and showed that its properties are closely connected with the prime-number distribution. He obtained various results and sketched the path of future progress in a number of well-founded conjectures of which all, except one that still remains undecided, have been shown to be correct. On the basis of Riemann's ideas, the prime-number theorem was proved independently in 1896 by the French mathematician J. Hadamard (1865–   ) and the Belgian C. J. de la Vallée-Poussin (1866–   ). Much progress has been made in analytic number theory since this time, but it remains a peculiar fact that the key to some of the most essential problems lies in the so-called Riemann's hypothesis, the last of his conjectures about the zeta function, which has not been demonstrated. It states that the complex zeros of the function all have the real component $\frac{1}{2}$.

Let us present another important result regarding the distribution of primes. As an example, the sequence of numbers

$$3, 7, 11, 15, 19, 23, 27, \ldots \qquad (4\text{--}21)$$

form an arithmetic series, *i.e.*, consecutive terms in the sequence have the same difference; in this case it is equal to 4. The general term in the sequence (4–21) is

$$4n - 1, \qquad n = 1, 2, 3, \cdots$$

The question arises whether this sequence contains an infinite number of primes. To see that this is true, one can apply a method that is a simple generalization of Euclid's idea for proving that there is an infinite number of primes. The assumption that there is only a finite number of primes

$$p_1 = 3, \qquad p_2, \cdots, p_k$$

in the sequence (4–21) leads to a contradiction, as we shall see. One could then form the new number

$$N = 4p_1 p_2 \cdots p_k - 1 = 4P - 1$$

which is not divisible by any p_i. But any odd prime is of one of the forms $4n + 1$ or $4n - 1$, and the product of two numbers of the form $4n + 1$ is again of this form, so at least one of the prime factors p of N is of the form $4n - 1$. But this prime cannot be any of the p_i's since they do not divide N; hence p is a new prime in the sequence (4–21).

The same argument may be used to show that the arithmetic series with the general term $6n - 1$, that is,

$$5, 11, 17, 23, 29, 35, \ldots$$

contains an infinite number of primes. In general, an arithmetic series consists of terms

$$an + b, \qquad n = 1, 2, \cdots \qquad (4\text{–}22)$$

where a and b are fixed numbers. If a and b have the common divisor d, all numbers in the sequence are divisible by d. But when one assumes that a and b are relatively prime, it can be shown that the sequence contains an infinite number of primes. This result is known as the theorem of Lejeune-Dirichlet (1805–1859). It is another of the many theorems in number theory that are simple to state and difficult to prove. Dirichlet's method requires complicated mathematical tools and many results from other fields. It is puzzling that many special cases can be obtained very simply, as we illustrated above, yet the search for an elementary proof has so far been unavailing.

Dirichlet's theorem states that the expressions (4–22) gives an infinite number of primes when $(a, b) = 1$. Attempts to show that other functions may have the same property have not succeeded. It has not even been possible to prove that an expression as simple as $n^2 + 1$ gives an infinite number of prime values. Related to these questions is the search for functions that will take only prime values for $n = 1, 2, \ldots$. We have already mentioned Fermat's unsuccessful conjecture. The other results on this problem are also all in the negative direction; one can show that certain types of functions cannot have the desired property.[1]

For instance, let us show that no polynomial with integral coefficients

$$f(x) = a_0 x^r + a_1 x^{r-1} + \cdots + a_{r-1} x + a_r$$

can take only prime values for integral x. Let us assume that for some $x = n$, the value $f(n) = p$ is a prime. Then for any integral t, the numbers $f(n + tp)$ are divisible by p, since

$$f(n + tp) - f(n) = a_0[(n + tp)^r - n^r] + a_1[(n + tp)^{r-1} - n^{r-1}]$$

$$+ \cdots + a_{r-1}[(n + tp) - n]$$

Also each difference

$$(n + tp)^i - n^i$$

is divisible by p, as one sees by the binomial expansion. Since every number $f(n + tp)$ is divisible by p, these numbers are composite unless

$$f(n + tp) = \pm p \qquad \text{or} \qquad f(n + tp) = 0 \qquad (4\text{--}23)$$

---

[1] W. H. Mills (*Bulletin American Mathematical Society*, June, 1947) has recently shown that there exists some real number A such that $[A^{3^n}]$ gives only primes. (The bracket denotes greatest integer as before.)

But a polynomial of degree r cannot take the same value more than r times so that the cases in (4–23) cannot happen for more than $3r$ values of t, at most, and for all other values $f(n + tp)$ must be composite.

Example.

When

$$f(x) = x^2 + 2x + 3$$

one finds $f(2) = 11$ and

$$f(2 + 11t) = 11(1 + 6t + 11t^2)$$

is composite and divisible by 11 when $t \neq 0$.

In connection with the prime values that polynomials will take, let us mention some peculiar examples of polynomials that take prime values for a long series of consecutive values of the variable. One is the polynomial

$$x^2 - x + 41$$

which produces a prime for the 41 values of x: 0, 1, 2, . . . , 40. Similarly

$$x^2 - 79x + 1{,}601$$

gives 80 consecutive prime values when $x = 0, 1, \cdots, 79$. There exist other examples of the same nature.

Let us conclude this review of facts and problems from the prime-number theory by a few remarks regarding the additive representation of numbers by means of primes. We have already mentioned the extensive correspondence between Euler and Goldbach regarding mathematical questions, particularly number theory. In some of these letters, dating from about 1742, Goldbach discusses the following two conjectures: Every even number ≥ 6 is the sum of two odd primes. Every odd number ≥ 9 is the sum of three odd primes. Euler, whose mathematical intuition was acute, states in reply that he also is convinced of the truth of these propositions, but he is unable to find any proof.

FACTOR TABLE

	1	3	7	9	11	13	17	19	21	23	27	29	31	33	37	39	41	43	47	49
0	..	..	..	3	..	..	..	..	3	..	3	..	..	3	..	3	..	..	..	7
100	..	..	..	..	3	..	3	7	11	3	..	3	..	7	..	..	3	11	3	..
200	3	7	3	11	..	3	7	3	13	..	..	..	3	..	3	..	..	3	13	3
300	7	3	..	3	..	..	..	11	3	17	3	7	..	3	..	3	11	7	..	..
400	..	13	11	..	3	7	3	..	..	3	7	3	..	..	19	..	3	..	3	..
500	3	..	3	..	7	3	11	3	..	..	17	23	3	13	3	7	..	3	..	3
600	..	3	..	3	13	..	..	..	3	7	3	17	..	3	7	3	..	..	..	11
700	..	19	7	..	3	23	3	..	7	3	..	3	17	..	11	..	3	..	3	7
800	3	11	3	..	..	3	19	3	..	..	..	..	3	7	3	..	29	3	7	3
900	17	3	..	3	..	11	7	..	3	13	3	..	7	3	..	3	..	23	..	13
1,000	7	17	19	..	3	..	3	..	..	3	13	3	..	..	17	..	3	7	3	..
1,100	3	..	3	..	11	3	..	3	19	..	7	..	3	11	3	17	7	3	31	3
1,200	..	3	17	3	7	..	..	23	3	..	3	..	..	3	..	3	17	11	29	..
1,300	..	..	..	7	3	13	3	..	..	3	..	3	11	31	7	13	3	17	3	19
1,400	3	23	3	..	17	3	13	3	7	..	..	..	..	3	..	..	11	3	..	3
1,500	19	3	11	3	..	17	37	7	3	..	3	11	..	3	29	3	23	..	7	..
1,600	..	7	..	..	3	..	3	..	..	3	..	3	7	23	..	11	3	31	3	17
1,700	3	13	3	..	29	3	17	3	..	..	11	7	3	..	3	37	..	3	..	3
1,800	..	3	13	3	..	7	23	17	3	..	3	31	..	3	11	3	7	19	..	43
1,900	..	11	..	23	3	..	3	19	17	3	41	3	..	..	..	13	7	3	29	3
2,000	3	..	3	7	..	3	..	3	43	7	..	..	3	19	3	..	13	3	23	3
2,100	11	3	7	3	..	..	29	13	3	11	3	..	..	3	..	3	..	..	19	7
2,200	31	..	3	47	3	..	3	7	..	3	17	3	23	7	..	3	..	3	3	13
2,300	3	7	3	..	..	3	7	3	11	23	13	17	3	..	3	..	3	..	3	..
2,400	7	3	29	3	..	19	..	41	3	..	3	7	11	3	..	3	..	7	..	31
2,500	41	..	23	13	3	7	3	11	..	3	7	3	..	17	43	..	3	..	3	..
2,600	3	19	3	..	7	3	..	3	..	43	37	11	3	..	3	7	19	3	..	3
2,700	37	3	..	3	..	..	11	..	3	7	3	..	..	3	7	3	..	13	41	..
2,800	..	..	7	53	3	29	3	..	7	3	11	3	19	..	..	17	3	..	3	7
2,900	3	..	3	..	41	3	..	3	23	37	..	29	3	7	3	..	17	3	7	3
3,000	..	3	31	3	..	23	7	..	3	..	3	13	7	3	..	3	..	17	11	..
3,100	7	29	13	..	3	11	3	..	..	3	53	3	31	13	..	43	3	7	3	47
3,200	3	..	3	..	13	3	..	3	..	11	7	..	3	53	3	41	7	3	17	3
3,300	..	3	..	3	7	..	31	..	3	..	3	..	..	3	47	3	13	..	..	17
3,400	19	41	..	7	3	..	3	13	11	3	23	3	47	..	7	19	3	11	3	..
3,500	3	31	3	11	..	3	..	3	7	13	..	..	3	..	3	..	..	3	..	3
3,600	13	3	..	3	23	..	..	7	3	..	3	19	..	3	..	3	11	..	7	41
3,700	..	7	11	..	3	47	3	..	61	3	..	3	7	..	37	..	3	19	3	23
3,800	3	..	3	13	37	3	11	3	..	..	43	7	3	..	3	11	23	3	..	3
3,900	47	3	..	3	..	7	..	..	3	..	3	..	..	3	31	3	7	..	..	11
4,000	..	..	..	19	3	..	3	..	..	3	..	3	29	37	11	7	3	13	3	..
4,100	3	11	3	7	..	3	23	3	13	7	..	..	3	..	3	..	41	3	11	3
4,200	..	3	7	3	11	..	..	3	41	3	..	..	3	19	3	..	..	..	31	7
4,300	11	13	59	31	3	19	3	7	29	3	..	3	61	7	..	..	3	43	3	..
4,400	3	7	3	..	11	3	7	3	..	..	19	43	3	11	3	23	..	3	..	3
4,500	7	3	..	3	13	..	..	..	3	..	3	7	23	3	13	3	19	7	..	..
4,600	43	..	17	11	3	7	3	31	..	3	7	3	11	41	..	..	3	..	3	..
4,700	3	..	3	17	7	3	53	3	..	..	29	..	3	..	3	7	11	3	47	3
4,800	..	3	11	3	17	..	..	61	3	7	3	11	..	23	7	3	..	29	37	13
4,900	13	..	7	..	3	17	3	..	7	3	13	3	..	..	..	11	3	..	3	7

Factor Table — (*Continued*)

	51	53	57	59	61	63	67	69	71	73	77	79	81	83	87	89	91	93	97	99
0	3	..	3	..	..	3	..	3	..	..	7	..	3	..	3	..	7	3	..	3
100	..	3	..	3	7	..	..	13	3	..	3	..	..	3	11	3	..	..	..	..
200	..	11	..	7	3	..	3	..	..	3	..	3	..	..	7	17	3	..	3	13
300	3	..	3	..	19	3	..	3	7	..	13	..	3	..	3	..	17	3	..	3
400	11	3	..	3	..	..	..	7	3	11	3	..	13	3	..	3	..	17	7	..
500	19	7	..	13	3	..	3	..	..	3	..	3	7	11	..	19	3	..	3	..
600	3	..	3	..	..	3	23	3	11	..	..	7	3	..	3	13	..	3	17	3
700	..	3	..	3	..	7	13	..	3	..	3	19	11	3	..	3	7	13	..	17
800	23	..	..	..	3	..	3	11	13	3	..	3	..	..	..	7	3	19	3	29
900	3	..	3	7	31	3	..	3	..	7	..	11	3	..	3	23	..	3	..	3
1,000	..	3	7	3	..	..	11	..	3	29	3	13	23	3	..	3	..	..	..	7
1,100	..	..	13	19	3	..	3	7	..	3	11	3	..	7	..	29	3	..	3	11
1,200	3	7	3	..	13	3	7	3	31	19	..	..	3	..	3	..	..	3	..	3
1,300	7	3	23	3	..	29	..	37	3	..	3	7	..	3	19	3	13	7	11	..
1,400	..	..	31	..	3	7	3	13	..	3	7	3	..	..	..	..	3	..	3	..
1,500	3	..	3	..	7	3	..	3	..	11	19	..	3	..	3	7	37	3	..	3
1,600	13	3	..	3	11	..	..	..	3	7	3	23	41	3	7	3	19	..	..	..
1,700	17	..	7	..	3	41	3	29	7	3	..	3	13	..	..	..	3	11	3	7
1,800	3	17	3	11	..	3	..	3	..	..	..	3	7	3	..	3	31	3	7	3
1,900	..	3	19	3	37	13	7	11	3	..	3	..	7	3	..	3	11	..	..	..
2,000	7	..	11	29	3	..	3	..	19	3	31	3	..	..	..	..	3	7	3	..
2,100	3	..	3	17	..	3	11	3	13	41	7	..	3	37	3	11	7	3	13	3
2,200	..	3	37	3	7	31	..	..	3	..	3	43	..	3	..	3	29	..	..	11
2,300	..	13	..	7	3	17	3	23	..	3	..	3	..	..	7	..	3	..	3	..
2,400	3	11	3	..	23	3	..	3	7	..	..	37	3	13	3	19	47	3	11	3
2,500	..	3	..	3	13	11	17	7	3	31	3	..	29	3	13	3	..	..	7	23
2,600	11	7	..	..	3	..	3	17	..	3	..	3	7	..	..	..	3	..	3	..
2,700	3	..	3	31	11	3	..	3	17	47	..	7	3	11	3	..	3	..	..	3
2,800	..	3	..	3	..	7	47	19	3	13	3	..	43	3	..	3	7	11	..	13
2,900	13	..	..	11	3	..	3	..	..	3	13	3	11	19	29	7	3	41	3	..
3,000	3	43	3	7	..	3	..	3	37	7	17	..	3	..	3	..	11	3	19	3
3,100	23	3	7	3	29	..	..	..	3	19	3	11	..	3	..	3	..	31	23	7
3,200	..	..	..	..	3	13	3	7	..	3	29	3	17	7	19	11	3	37	3	..
3,300	3	7	3	..	..	3	7	3	..	..	11	31	3	17	3	..	..	3	43	3
3,400	7	3	..	3	..	..	..	..	3	23	3	7	59	3	11	3	..	7	13	..
3,500	53	11	..	..	3	7	3	43	..	3	7	3	..	..	17	37	3	..	3	59
3,600	3	13	3	..	7	3	19	3	..	..	..	13	3	29	3	7	..	3	..	3
3,700	11	3	13	3	..	53	..	..	3	7	3	..	19	3	7	3	17	..	..	29
3,800	..	..	7	17	3	..	3	53	7	3	..	3	..	11	13	..	3	17	3	7
3,900	3	59	3	37	17	3	..	3	11	29	41	23	3	7	3	..	13	3	7	3
4,000	..	3	..	3	31	17	7	13	3	..	3	..	7	3	61	3	..	..	17	..
4,100	7	..	..	..	3	23	3	11	43	3	..	3	37	47	53	59	3	7	3	13
4,200	3	..	3	..	..	3	17	3	..	..	7	11	3	..	3	..	7	3	..	3
4,300	19	3	..	3	7	..	11	17	3	..	3	29	13	3	41	3	..	23	..	53
4,400	..	61	..	7	3	..	3	41	17	3	11	3	..	..	7	67	3	..	3	11
4,500	3	29	3	47	..	3	..	3	7	17	23	19	3	..	3	13	..	3	..	3
4,600	..	3	..	3	59	..	13	7	3	..	3	..	31	3	43	3	..	13	7	37
4,700	..	7	67	..	3	11	3	19	13	3	17	3	7	..	..	..	3	..	3	..
4,800	3	23	3	43	..	3	31	3	..	11	..	7	3	19	3	..	67	3	59	3
4,900	..	3	..	3	11	7	..	..	3	..	3	13	17	3	..	3	7	..	19	..

One verifies Goldbach's conjectures immediately for the smallest numbers.   For instance, for even numbers,

$$6 = 3 + 3 \qquad\qquad 14 = 3 + 11$$
$$8 = 3 + 5 \qquad\qquad 16 = 3 + 13$$
$$10 = 3 + 7 \qquad\qquad 18 = 5 + 13$$
$$12 = 5 + 7 \qquad\qquad 20 = 3 + 17$$

and for odd numbers,

$$9 = 3 + 3 + 3 \qquad\qquad 17 = 3 + 7 + 7$$
$$11 = 3 + 3 + 5 \qquad\qquad 19 = 3 + 5 + 11$$
$$13 = 3 + 3 + 7 \qquad\qquad 21 = 3 + 5 + 13$$
$$15 = 3 + 5 + 7$$

The smallest integers 1, 2, 3, 5 must obviously be regarded as exceptions.   When the numbers become fairly large, there will usually be numerous representations; for instance,

$$48 = 5 + 43 = 7 + 41 = 11 + 37 = 17 + 31 = 19 + 29$$

Goldbach's conjectures have been verified numerically up to 100,000 (N. Pipping).   One should note also that the first conjecture implies the second.   Take an odd number N and subtract the odd prime $p < N$ from it.   Then $N - p$ is even, and if the even numbers could be expressed as the sum of two primes, any odd number N would be the sum of three.

A problem for which Euler could find no attacking point could be expected to be extremely difficult, and it was not until fairly recently that essential progress was made.   The Norwegian mathematician V. Brun (1885–   ) developed an extension of the sieve method of Eratosthenes that enabled him to show that every sufficiently large even number N can be written as a sum

$$N = N_1 + N_2$$

where N_1 and N_2 have at most nine prime factors.   Later, others improved the result to four prime factors, but it is still a far cry to

Goldbach's conjecture, which requires a single prime factor in each summand. Goldbach's second theorem, which we saw was a weaker result that would follow from the first, is, however, much nearer to its final solution. In 1937 the Russian mathematician I. Vinogradoff succeeded in showing by analytic means that every odd number that is sufficiently large is the sum of three odd primes. How large the numbers have to be, however, he could not decide.

Bibliography

CUNNINGHAM, A. J. C.: *Binomial Factorisations Giving Extensive Congruence-Tables and Factorisation-Tables*, Vols. 1–7, Francis Hodgson, London, 1923–25.

KRAITCHIK, M.: *Théorie des nombres,* Vols. I and II, Gauthier-Villars & Cie, Paris, 1922, 1926.

LEHMER, D.N.: "Factor table for the first ten millions containing the smallest factor of every number not divisible by 2, 3, 5 and 7 between the limits 0 and 10,017,000," *Carnegie Institution of Washington Publication* 105 (1909).

————: "List of Prime Numbers from 1 to 10,006,721," *Carnegie Institution of Washington Publication* 165 (1914).

CHAPTER 5

THE ALIQUOT PARTS

5–1. The divisors of a number. Several problems relating to the divisors of a number can be solved by means of the main theorem that every integer can be represented uniquely as the product of prime factors. A number N shall be written

$$N = p_1^{\alpha_1} p_2^{\alpha_2} \cdots p_r^{\alpha_r} \tag{5-1}$$

where the p_i's are the various different prime factors and α_i the *multiplicity*, i.e., the number of times p_i occurs in the prime factorization. For any divisor d of N one has

$$N = dd_1 \tag{5-2}$$

where d_1 is the divisor paired with d. When multiplied together, the prime factorizations of d and d_1 must give that of N so that

$$d = p_1^{\delta_1} p_2^{\delta_2} \cdots p_r^{\delta_r} \tag{5-3}$$

where the exponents δ_i do not exceed the corresponding α_i in (5–1). Since the second factor in (5–2) must contain the remaining factors, it becomes

$$d_1 = p_1^{\alpha_1-\delta_1} p_2^{\alpha_2-\delta_2} \cdots p_r^{\alpha_r-\delta_r}$$

In the expression (5–3) for a divisor the exponent δ_1 can take the $\alpha_1 + 1$ values $0, 1, \ldots, \alpha_1$, similarly δ_2 the $\alpha_2 + 1$ values $0, 1, \ldots, \alpha_2$, and so on. Since each choice of δ_1 can be combined with any choice of δ_2, and so on, one concludes:

Theorem 5–1. The number of divisors of a number N in the form (5–1) is

$$\nu(N) = (\alpha_1 + 1)(\alpha_2 + 1) \cdots (\alpha_r + 1) \tag{5-4}$$

Example.

The number

$$60 = 2^2 \cdot 3 \cdot 5$$

has

$$\nu(60) = (2 + 1)(1 + 1)(1 + 1) = 12$$

divisors. They are

$$1, \quad 2, \quad 3, \quad 4, \quad 5, \quad 6, \quad 10, \quad 12, \quad 15, \quad 20, \quad 30, \quad 60 \qquad (5\text{–}5)$$

We shall now determine various expressions that may be formed by means of the divisors of a number. We find first the *product* of the divisors. In (5–2) let d run through all $\nu(N)$ divisors of N. The corresponding d_1 will then also run through these divisors in some order, so that the product of all d's is the same as the product of all d_1's. This we write

$$\Pi d = \Pi d_1$$

where the symbol Π, as usual, denotes the product. We form the product of all $\nu(N)$ equations (5–2) and obtain

$$N^{\nu(N)} = (\Pi d)(\Pi d_1) = (\Pi d)^2$$

This yields the desired result:

THEOREM 5–2. The product of all divisors of a number N is

$$\Pi d = N^{\frac{1}{2}\nu(N)}$$

Example.

The product of all divisors of 60 is

$$\Pi d = 60^6 = 46{,}656{,}000{,}000$$

as one may check by multiplying together the divisors (5–5).

The result in theorem 5–2 can be expressed in a different manner. When

$$x_1, \ x_2, \ \ldots, \ x_n \qquad (5\text{–}6)$$

is a set of n positive numbers, the *geometric mean* of the numbers is defined to be

$$G = \sqrt[n]{x_1 x_2 \cdots x_n}$$

When applied to the product of the $\nu(N)$ divisors of N, one sees:

THEOREM 5–3.   The geometric mean of the divisors of a number N is

$$G(N) = \sqrt{N}$$

The determination of the sum $\sigma(N)$ of the divisors of a number N is slightly more complicated.   We shall first reduce the problem to the case where N is a power of a single prime, a method that is often applicable in similar problems.   Let us write the given number N as a product of two relatively prime factors

$$N = ab$$

Since the prime factors of a and b are different, one concludes that any divisor d of N must have the form

$$d = a_i b_i \tag{5–7}$$

where a_i is a divisor of a and b_i a divisor of b.   We denote the divisors of a and b, respectively, by

$$1, a_1, a_2, \ldots, a, \qquad 1, b_1, b_2, \ldots, b$$

so that their sums are

$$\sigma(a) = 1 + a_1 + a_2 + \cdots + a, \qquad \sigma(b) = 1 + b_1 + b_2 + \cdots + b$$

In (5–7) let us take all divisors of N with the same a_i.   Their sum is

$$a_i(1 + b_1 + b_2 + \cdots + b) = a_i \sigma(b)$$

Next, by taking this sum for all possible a_i one obtains as the total sum of all divisors of N

$$1\sigma(b) + a_1\sigma(b) + \cdots + a\sigma(b) = \sigma(a)\sigma(b)$$

Thus we have derived the result that when a and b are relatively prime

$$\sigma(N) = \sigma(ab) = \sigma(a)\sigma(b) \tag{5–8}$$

We split a and b further into relatively prime factors and apply the same rule (5–8) again.   This may be continued until the factors become the powers of the various primes dividing N.   As a consequence we conclude that when N has the prime factorization (5–1)

$$\sigma(N) = \sigma(p_1^{\alpha_1})\sigma(p_2^{\alpha_2}) \cdots \sigma(p_r^{\alpha_r}) \qquad (5\text{–}9)$$

For a prime power p^α the divisors are

$$1, p, p^2, \ldots, p^\alpha$$

so that

$$\sigma(p^\alpha) = 1 + p + p^2 + \cdots + p^\alpha$$

This is a geometric series in which the quotient of two consecutive terms is p. It may be summed by the usual trick, multiplying the series by p

$$p \cdot \sigma(p^\alpha) = p + p^2 + \cdots + p^\alpha + p^{\alpha+1}$$

and subtracting the original series

$$p \cdot \sigma(p^\alpha) - \sigma(p^\alpha) = p^{\alpha+1} - 1$$

so that

$$\sigma(p^\alpha) = \frac{p^{\alpha+1} - 1}{p - 1} \qquad (5\text{–}10)$$

When this result is applied to each factor in (5–9), we have proved:

THEOREM 5–4. The sum of divisors of a number N with the prime factorization (5–1) is

$$\sigma(N) = \frac{p_1^{\alpha_1+1} - 1}{p_1 - 1} \cdot \frac{p_2^{\alpha_2+1} - 1}{p_2 - 1} \cdot \ldots \cdot \frac{p_r^{\alpha_r+1} - 1}{p_r - 1} \qquad (5\text{–}11)$$

Each prime power p^α in the factorization (5–1) contributes a factor (5–10) to the expression (5–11) for the sum of the divisors. It is useful to observe the two following simple cases, which occur commonly. When there is a single prime factor p, one has

$$\sigma(p) = \frac{p^2 - 1}{p - 1} = p + 1$$

When $p = 2^\alpha$,

$$\sigma(2^\alpha) = \frac{2^{\alpha+1} - 1}{2 - 1} = 2^{\alpha+1} - 1$$

Let us mention also that the *average* or *arithmetic mean* $A(N)$ of the divisors is obtained by dividing their sum $\sigma(N)$ by their number $\nu(N)$ so that

$$A(N) = \frac{\sigma(N)}{\nu(N)} \qquad (5\text{-}12)$$

Example.

The sum of the divisors of

$$60 = 2^2 \cdot 3 \cdot 5$$

is

$$\sigma(60) = (2^3 - 1)(3 + 1)(5 + 1) = 168$$

as one can verify by summing the divisors (5–5). Their average is

$$A(60) = \frac{168}{12} = 14$$

Since $14 > \sqrt{60}$ this illustrates the general fact that the arithmetic mean is greater than the geometric mean.

The *harmonic mean* H of a set of numbers (5–6) is defined by

$$\frac{1}{H} = \frac{1}{n} \cdot \left(\frac{1}{x_1} + \frac{1}{x_2} + \cdots + \frac{1}{x_n} \right) \qquad (5\text{-}13)$$

To determine the harmonic mean $H(N)$ of the divisors of a number N let us first find the sum of their inverse values. According to (5–2) one has for any divisor d

$$\frac{1}{d} = \frac{d_1}{N}$$

where d_1 is the divisor paired with d. Here, as before, when d runs through all divisors of N, so will d_1. By summing all these equations for the various d's, one obtains therefore

$$\sum \frac{1}{d} = \frac{1}{N} \sum d_1 = \frac{1}{N} \sum d = \frac{\sigma(N)}{N}$$

where $\sum$ is the usual summation symbol. According to (5–13), we divide by the number $\nu(N)$ of divisors and find by means of (5–12)

$$\frac{1}{H(N)} = \frac{1}{\nu(N)} \sum \frac{1}{d} = \frac{\sigma(N)}{\nu(N)N} = \frac{A(N)}{N}$$

This gives the result:

THEOREM 5–5. The product of the harmonic and arithmetic mean of the divisors of a number is equal to the number itself

$$N = A(N) \cdot H(N)$$

Since the arithmetic mean is greater than the geometric mean $\sqrt{N}$ of the divisors, one has the inequality

$$A(N) \geqq \sqrt{N} \geqq H(N)$$

in accordance with the general theory of means.

Example.

We have seen that the arithmetic mean of the divisors of 60 is 14. Therefore

$$H(60) = \tfrac{60}{14} = 4\tfrac{2}{7}$$

Problems.

1. Find the number of divisors, their sum, and their means for (a) 220, (b) 365, (c) $6! = 1 \cdot 2 \cdot 3 \cdot 4 \cdot 5 \cdot 6$.

2. Find the sum of the squares of the divisors of a number.

3. Find the smallest numbers with 2, 3, 4, 5, or 6 divisors.

5–2. Perfect numbers. The perfect numbers are essential elements in all numerological speculations. God created the world in six days, a perfect number. The moon circles the earth in 28 days, again a symbol of perfection in the best of all possible worlds. In numerological terminology, the divisors are the *parts* of which a number is created or reproduced. A *perfect number* is a number that is the sum of its divisors, or, in more archaic language, it is the sum of its *aliquot parts*. In this definition it must be observed that Greek mathematics excluded the number itself as a proper part. Therefore, to obtain the sum $\sigma_0(N)$ of the aliquot parts of a number N in the Greek sense, one must diminish the sum $\sigma(N)$ of all divisors we found in theorem 5–4 by the improper divisor N so that

$$\sigma_0(N) = \sigma(N) - N \qquad (5\text{--}14)$$

The condition for a perfect number may then be expressed in the formula

$$\sigma_0(N) = N \qquad (5\text{--}15)$$

or equivalently

$$\sigma(N) = 2N \qquad (5\text{--}16)$$

By means of any of these conditions one can check whether a number is perfect. For instance,

$$\sigma(6) = \sigma(2 \cdot 3) = (2 + 1)(3 + 1) = 12$$

and

$$\sigma(28) = \sigma(2^2 \cdot 7) = (2^3 - 1)(7 + 1) = 56$$

so that both 6 and 28 are perfect numbers.

Only one general type of perfect numbers is known:

THEOREM 5–6. A number of the form

$$P = 2^{p-1}(2^p - 1) \qquad (5\text{--}17)$$

is perfect when

$$q = 2^p - 1$$

is a Mersenne prime.

This theorem represents the final proposition in the ninth book of Euclid's *Elements*. The proof consists in computing

$$\sigma(P) = (2^p - 1)(q + 1) = (2^p - 1)2^p = 2P$$

We have already mentioned that there are only 12 known Mersenne primes, which one obtains for

$$p = 2, 3, 5, 7, 13, 17, 19, 31, 61, 89, 107, 127$$

From these one computes the 12 known perfect numbers. The first four are

$$P_2 = 2 \cdot (2^2 - 1) = 6$$
$$P_3 = 2^2 \cdot (2^3 - 1) = 28$$
$$P_5 = 2^4 \cdot (2^5 - 1) = 496$$
$$P_7 = 2^6 \cdot (2^7 - 1) = 8{,}128$$

In Barlow's *Number Theory* (London, 1811) the author gives the perfect numbers up to P_{31} corresponding to the Mersenne prime M_{31} obtained by Euler, at the time the greatest prime known. This perfect number "is the greatest that will ever be discovered, for, as they are merely curious without being useful it is not likely that any person will attempt to find one beyond it." The great efforts expended since that time in such computations show that it is difficult to underestimate human curiosity. At present it seems possible that further efforts along these lines will be made by means of the tremendous calculators developed during the Second World War, as soon as they are available for more peaceful pursuits.

All of these perfect numbers are even; Euler succeeded in proving the following theorem, which is the most general result known for perfect numbers:

THEOREM 5–7. Every even perfect number is of the type (5–17) discussed by Euclid.

To prove this theorem we write the even perfect number P in the form

$$P = 2^{p-1} \cdot q \tag{5–18}$$

where q is some odd number. Since the two factors in (5–18) are relatively prime, one finds as in (5–8)

$$\sigma(P) = \sigma(2^{p-1}) \cdot \sigma(q) = (2^p - 1) \cdot \sigma(q)$$

The condition (5–16) for a perfect number states that one must have

$$\sigma(P) = (2^p - 1)\sigma(q) = 2P = 2^p \cdot q$$

In this relation let us use the sum $\sigma_0(q)$ of the proper divisors as defined in (5–14) instead of $\sigma(q)$. One obtains

$$(2^p - 1)[\sigma_0(q) + q] = 2^p q$$

and this may be rewritten

$$q = (2^p - 1)\sigma_0(q) \tag{5–19}$$

This condition permits us to draw some strong conclusions. It implies first that $d = \sigma_0(q)$ is a proper divisor of q. On the other hand $\sigma_0(q)$ was the sum of *all* proper divisors of q, including d, so that there cannot be any other proper divisors besides d. But a number q with a single proper divisor d must be a prime and $d = 1$. From (5–19) one concludes finally that

$$q = 2^p - 1$$

is a Mersenne prime. Thus the even perfect number (5–18) is of the form (5–17) given by Euclid.

Do there exist any odd perfect numbers? This is one of the celebrated unsolved problems in number theory. Extensive numerical computations have failed to divulge any odd perfect number less than 2,000,000. It has been possible to find various conditions that such numbers must satisfy but they are insufficient to prove that odd perfect numbers cannot exist. (See Supplement.)

For numbers that are not perfect there are the two possibilities:

$$\sigma_0(N) > N, \qquad \sigma_0(N) < N$$

Numbers of the first kind are called *abundant*, and those of the second kind are *deficient*. This distinction is considered important in numerology. For instance, Alcuin (735–804), the adviser and teacher of Charlemagne, observes that the entire human race descends from the 8 souls in Noah's ark. Since 8 is a deficient number, he concludes that this second creation was imperfect in comparison with the first, which was based on the principle of the perfect number 6. The perfect numbers represent the happy medium between abundance and deficiency.

The first few abundant numbers are

$$12, 18, 20, 24, 30, 36, \ldots$$

There are only 21 abundant numbers up to 100, as the reader may verify, and they are all even. The first odd abundant number is

$$945 = 3^3 \cdot 5 \cdot 7$$

for which

$$\sigma_0(945) = 975$$

There exists a table of the values of the sum $\sigma(N)$ of the divisors of the numbers up to 10,000, computed by J. W. L. Glaisher, so that the character of numbers not exceeding this limit is easily determined. There are some rules for abundant and deficient numbers: for instance, a prime or a power of a prime is deficient; any divisor of a perfect or deficient number is deficient; any multiple of an abundant or perfect number is abundant.

We saw that the perfect numbers were defined by the condition

$$\sigma_0(N) = N$$

For certain abundant numbers the sum of the proper divisors may turn out to be a multiple of the number itself. For example

$$\sigma_0(120) = 2 \cdot 120$$

as one easily verifies. A number of this kind is called *multiply perfect*. When

$$\sigma_0(N) = k \cdot N$$

the integer k may be called the *class* of the multiply perfect number, so that 120 is of class 2 while the perfect numbers are of class 1.

The problem of finding such multiply perfect numbers appears to have been formulated first in 1631 by Mersenne in a letter to René Descartes (1596–1650). Although Descartes's fame rests mainly on his philosophical method and in mathematics on his creation of analytical geometry and the invention of coordinate systems, he was also greatly interested in number theory and made various contributions to it. He must have speculated considerably over the problem proposed by Mersenne, because about seven years later he responded with a list of multiply perfect numbers, which he could not have discovered without great effort and ingenuity. In the meanwhile, Fermat had also tackled the problem and discovered a second multiply perfect number, namely,

$$672 = 2^5 \cdot 3 \cdot 7$$

for which

$$\sigma_0(672) = 2 \cdot 672$$

while André Jumeau, prior of Sainte-Croix, found a third

$$523,776 = 2^9 \cdot 3 \cdot 11 \cdot 31$$

also of class 2. Descartes in several letters to Mersenne gave another multiply perfect number of class 2, namely, 1,476,304,896, six others of class 3, and one of class 4. In addition he described

Fig. 5-1. René Descartes (1596-1650).

various general rules that permitted him to construct these numbers. The subsequent letters exchanged between Mersenne, Fermat, and Frénicle contain several other multiply perfect numbers. More recently many others have been discovered, notably by E. Lucas, D. N. Lehmer, A. Cunningham, R. D. Carmichael, and D. E. Mason. The most complete list to date is due to P. Poulet (1929), and it contains 334 multiply perfect numbers, some of class as high as 7. (See Supplement.)

5-3. Amicable numbers. Another type of numbers that are prominent in the lore of number mysticism is the *amicable numbers*. They are defined to be pairs of numbers such that each member is composed of the parts of the other, thus symbolizing mutual harmony, perfect friendship, and love. The existence of amicable numbers seems to have been discovered somewhat later than the perfect numbers, probably in the period of the flowering of the Neo-Platonic mystical school in Greek philosophy. One of the most influential of the Neo-Platonic philosophers, Iamblichus of Chalcis (about A.D. 320), ascribes the knowledge of amicable numbers to the earliest Pythagorean school (about 500 B.C.). This mythical tradition has, however, little credit with the historians of the mathematical sciences.

In Arab mathematical writings the amicable numbers occur repeatedly. They play a role in magic and astrology, in the casting of horoscopes, in sorcery, in the concoction of love potions, and in the making of talismans. As an illustration let us quote from the *Historical Prolegomenon* of the Arab scholar Ibn Khaldun (1332–1406):

Let us mention that the practice of the art of talismans has also made us recognize the marvelous virtues of amicable (or sympathetic) numbers. These numbers are 220 and 284. One calls them amicable because the aliquot parts of one when added give a sum equal to the other. Persons who occupy themselves with talismans assure that these numbers have a particular influence in establishing union and friendship between two individuals. One prepares a horoscope theme for each individual, the first under the sign of Venus while this planet is in its house or in its exaltation and while it presents in regard to the moon an aspect of love and benevolence. In the second theme the ascendant should be in the seventh sign. On each one of these themes one inscribes one of the numbers just indicated, but giving the strongest number to the person whose friendship one wishes to gain, the beloved person. I don't know if by the strongest number one wishes to designate the greatest one or the one which has the greatest number of aliquot parts. There results a bond so close between the two persons that they cannot be separated. The author of the Ghaïa and other great masters in this art declare that they have seen this confirmed by experience.

Through the Arabs the knowledge of amicable numbers spread to Western Europe. They are mentioned in the works of many prominent mathematical writers around A.D. 1500, for instance Nicolas Chuquet, Etienne de la Roche, known as Villefranche, Michael Stiefel, Cardanus, and Tartaglia.

As we have already stated, a pair of numbers is said to be amicable when the sum of the aliquot parts of one is equal to the other, and conversely. In our previous terminology this can be expressed that M and N are amicable when

$$\sigma_0(N) = M, \qquad \sigma_0(M) = N \qquad (5\text{–}20)$$

When one uses the sums of all divisors of the numbers, the conditions (5–20) may be restated

$$\sigma(N) = \sigma(M) = N + M \qquad (5\text{–}21)$$

In ancient numerology there appears but a single set of amicable numbers, namely, the pair

$$N = 220 = 2^2 \cdot 5 \cdot 11, \qquad M = 284 = 2^2 \cdot 71$$

Even the discovery of the special relations between these two fairly large numbers is evidence of considerable familiarity with number properties. For this pair one has

$$M + N = 504$$

and the formula for the sum of the divisors of a number yields

$$\sigma(N) = (2^3 - 1)(5 + 1)(11 + 1) = 504$$
$$= (2^3 - 1)(71 + 1) = \sigma(M)$$

so that the condition (5–21) is fulfilled.

There is no indication of any other pair of amicable numbers having been discovered before the work of Fermat. This is somewhat peculiar since Fermat found his new pair through the rediscovery of a rule that actually had been formulated by the Arab mathematician Abu-l-Hasan Thabit ben Korrah as early as the ninth century. This rule we shall reformulate as follows:

For the various exponents n, write down in a table the numbers

$$p_n = 3 \cdot 2^n - 1 \qquad (5\text{–}22)$$

As may be seen, each number is obtained by doubling the preceding and adding 1.

n	1	2	3	4	5	6	7
p_n	5	11	23	47	95	191	383

(5–23)

If, for some n, two successive terms p_{n-1} and p_n are both primes, one examines the number

$$q_n = 9 \cdot 2^{2n-1} - 1 \qquad (5\text{–}24)$$

If this number is also prime the pair

$$M = 2^n p_{n-1} p_n, \qquad N = 2^n q_n \qquad (5\text{–}25)$$

is amicable. To illustrate the rule we observe that $p_1 = 5$ and $p_2 = 11$ are primes, and since $q_2 = 71$ is also prime we obtain the classical pair 220 and 284 from (5–25).

To prove the rule of Thabit ben Korrah we compute by means of (5–22) and (5–24)

$$\sigma(M) = (2^{n+1} - 1)(p_{n-1} + 1)(p_n + 1) = 9 \cdot 2^{2n-1}(2^{n+1} - 1)$$

$$\sigma(N) = (2^{n+1} - 1)(q_n + 1) \qquad\quad = 9 \cdot 2^{2n-1}(2^{n+1} - 1)$$

Since also

$$M + N = 2^n \cdot (p_{n-1}p_n + q_n) = 9 \cdot 2^{2n-1}(2^{n+1} - 1)$$

the pair is amicable.

The next pair of successive primes in the table (5–23) is $p_3 = 23$ and $p_4 = 47$. In this case $q_4 = 1,151$ is also prime, and we obtain the amicable pair announced by Fermat in 1636

$$17,296 = 2^4 \cdot 23 \cdot 47, \qquad 18,416 = 2^4 \cdot 1,151$$

Descartes stated in letters to Mersenne in 1638 that he had been led to the same rule and gave the third pair of amicable numbers

$$9,363,584 = 2^7 \cdot 191 \cdot 383, \qquad 9,437,056 = 2^7 \cdot 73,727$$

corresponding to the primes $p_6 = 191$ and $p_7 = 383$ in the series (5–23).

Euler took up the search for amicable numbers in a systematic manner and developed several methods for finding them.   In 1747 he gave a list of 30 pairs which he later expanded to more than 60. Some of them are

$$\begin{cases} 2 \cdot 5 \cdot 7 \cdot 19 \cdot 107 \\ 2 \cdot 5 \cdot 47 \cdot 359 \end{cases} \qquad \begin{cases} 2^4 \cdot 47 \cdot 89 \\ 2^4 \cdot 53 \cdot 79 \end{cases} \qquad \begin{cases} 2^3 \cdot 19 \cdot 41 \\ 2^5 \cdot 199 \end{cases}$$

$$\begin{cases} 2^4 \cdot 23 \cdot 479 \\ 2^4 \cdot 89 \cdot 127 \end{cases} \qquad \begin{cases} 2^2 \cdot 5 \cdot 251 \\ 2^2 \cdot 13 \cdot 107 \end{cases} \qquad \begin{cases} 2^3 \cdot 17 \cdot 79 \\ 2^3 \cdot 23 \cdot 59 \end{cases}$$

A rank amateur may occasionally make a contribution to number theory, as was demonstrated again when the sixteen-year-old Italian boy Nicolò Paganini in 1866 published the very small pair of amicable numbers

$$1{,}184 = 2^5 \cdot 37, \qquad 1{,}210 = 2 \cdot 5 \cdot 11^2$$

which had eluded all previous investigators.   They were probably found by trial and error.   An extensive list of amicable numbers is due to P. Poulet (1929).   A complete survey of the existing knowledge about amicable numbers has recently been published by E. B. Escott.   It contains a list of the 390 known amicable pairs together with the names of their discoverers.

5-4. Greatest common divisor and least common multiple. The algorism of Euclid enabled us to find the greatest common divisor of two and more numbers and also their least common multiple.   When the prime-factor decompositions of the numbers are known, the process becomes much simpler.   Let a and b be two given numbers, and

$$a = p_1^{\alpha_1} p_2^{\alpha_2} \cdots p_r^{\alpha_r}, \qquad b = p_1^{\beta_1} p_2^{\beta_2} \cdots p_r^{\beta_r} \qquad (5\text{--}26)$$

be their prime factorizations.   It is convenient to write the two decompositions formally as if the same primes occur in both. This is possible since, for instance, if p_1 should not divide b one can take $\beta_1 = 0$.   Since one is interested in what happens in regard to each prime p_i, it often simplifies matters to use the product symbol and write instead of (5–26)

$$a = \prod_i p_i^{\alpha_i}, \qquad b = \prod_i p_i^{\beta_i} \qquad (5\text{--}27)$$

When a number d is to divide both a and b, it cannot have any prime factors different from those occurring in these numbers so that one can write

$$d = p_1^{\delta_1} p_2^{\delta_2} \cdots p_r^{\delta_r} \qquad (5\text{--}28)$$

For each i the exponent δ_i in (5–28) cannot exceed any of the corresponding α_i and β_i in (5–26). If, therefore, d is to be the g.c.d. of a and b, the exponent δ_i must be the smaller or minimal of the two exponents α_i and β_i. In mathematical shorthand we write

$$\delta_i = \min (\alpha_i, \beta_i), \qquad i = 1, 2, \cdots, r$$

Similarly if

$$m = p_1^{\mu_1} p_2^{\mu_2} \cdots p_r^{\mu_r}$$

is to be divisible both by a and b, none of the exponents μ_i can be less than α_i or β_i. Therefore, if m is the l.c.m. of a and b, each μ_i must be equal to the greater or maximal of the two numbers α_i and β_i; in symbols

$$\mu_i = \max (\alpha_i, \beta_i), \qquad i = 1, 2, \cdots, r$$

Let us summarize these remarks:

THEOREM 5–8. The greatest common divisor and the least common multiplum of the two numbers a and b with the prime decompositions (5–27) are respectively

$$(a, b) = \prod p_i^{\min (\alpha_i, \beta_i)}, \qquad [a, b] = \prod_i p_i^{\max (\alpha_i, \beta_i)} \qquad (5\text{--}29)$$

It is evident that these rules (5–29) can be extended to three or an arbitrary set of numbers.

Example.

For

$$a = 2^6 \cdot 3^2 \cdot 5, \qquad b = 2^5 \cdot 3^3 \cdot 7$$

one has the g.c.d. and l.c.m.

$$(a, b) = 2^5 \cdot 3^2, \qquad [a, b] = 2^6 \cdot 3^3 \cdot 5 \cdot 7$$

Let us show how some of the properties of the g.c.d. and the l.c.m. we derived previously (Chap. 3) follow quite simply also from the formulas (5–29). If we multiply the two numbers (a, b) and $[a, b]$, the exponent to each p_i becomes

$$\min (\alpha_i, \beta_i) + \max (\alpha_i, \beta_i)$$

But the sum of the smaller and the greater of two numbers is their sum $\alpha_i + \beta_i$. On the other hand $\alpha_i + \beta_i$ is the exponent of p_i in the product ab so that we have (3–4)

$$(a, b)[a, b] = ab \tag{5–30}$$

Let us next multiply the g.c.d. (a, b) by a number c with the prime decomposition

$$c = p_1^{\gamma_1} p_2^{\gamma_2} \cdots p_r^{\gamma_r}$$

The exponent of p_i in the product $c \cdot (a, b)$ is then

$$\gamma_i + \min (\alpha_i, \beta_i)$$

But this is the same as the number

$$\min (\gamma_i + \alpha_i, \gamma_i + \beta_i)$$

which is the exponent of p_i in (ca, cb). This shows us that

$$c(a, b) = (ca, cb) \tag{5–31}$$

as we obtained previously in (3–2). Similarly one sees that for the l.c.m. (3–5)

$$c[a, b] = [ca, cb] \tag{5–32}$$

holds, because one has the identity

$$\gamma_i + \max (\alpha_i, \beta_i) = \max (\gamma_i + \alpha_i, \gamma_i + \beta_i)$$

The laws (5–30), (5–31), and (5–32) were derived here by means of the theorem of the unique factorization of a number into prime factors. It is of interest to note that each of them is the expression of some simple property of the process of forming maximum and minimum of two numbers, namely,

$$\min (\alpha, \beta) + \max (\alpha, \beta) = \alpha + \beta$$

$$\gamma + \min (\alpha, \beta) = \min (\gamma + \alpha, \gamma + \beta)$$

$$\gamma + \max (\alpha, \beta) = \max (\gamma + \alpha, \gamma + \beta)$$

A little later on we shall derive some properties of the g.c.d. and the l.c.m. that depend on the same principle, but involve the maximum and minimum of three numbers.

Let us discuss the two operations of forming the g.c.d. and the l.c.m. from a somewhat different point of view. Each of them

$$d = (a, b), \qquad m = [a, b]$$

associates new elements d and m with the given ones a and b, much in the way of ordinary addition and multiplication. These operations satisfy some very simple laws:

1. *Idempotent law:*

$$(a, a) = a, \qquad\qquad [a, a] = a$$

2. *Commutative law:*

$$(a, b) = (b, a), \qquad\qquad [a, b] = [b, a]$$

3. *Associative law:*

$$((a, b), c) = (a, (b, c)), \qquad [[a, b], c] = [a, [b, c]]$$

4. *Absorption law:*

$$[a, (a, b)] = a, \qquad\qquad (a, [a, b]) = a$$

Let us make a few comments on the four properties of the operations. First, it should be noticed that the two operations are dual, *i.e.*, the conditions remain the same when the g.c.d. and the l.c.m. are exchanged everywhere. The idempotent condition states only that the g.c.d. and the l.c.m. of a number a with itself is a. The commutative law and the associative law are exactly the same as for addition and multiplication

$$a + b = b + a, \qquad\qquad ab = ba$$

$$a + (b + c) = (a + b) + c, \qquad a(bc) = (ab)c$$

Since (a, b) is a divisor of a, the l.c.m. of a and (a, b) must be a and the second part of the absorption law is equally trivial.

Here also, the four laws can be expressed as properties of the operation of forming maximum and minimum of numbers. Let p be some prime that divides a, b, and c to the powers p^α, p^β, and p^γ, respectively. Then we leave it to the reader to verify that the four laws are consequences of

1.

$$\min(\alpha, \alpha) = \alpha, \qquad \max(\alpha, \alpha) = \alpha$$

2.

$$\min(\alpha, \beta) = \min(\beta, \alpha), \qquad \max(\alpha, \beta) = \max(\beta, \alpha)$$

3.

$$\min\{\min(\alpha, \beta), \gamma\} = \min\{\alpha, \min(\beta, \gamma)\}$$
$$\max\{\max(\alpha, \beta), \gamma\} = \max\{\alpha, \max(\beta, \gamma)\}$$

4.

$$\max\{\alpha, \min(\alpha, \beta)\} = \alpha, \qquad \min\{\alpha, \max(\alpha, \beta)\} = \alpha$$

One reason for going into these simple rules for the g.c.d. and the l.c.m. in some detail is that mathematicians quite recently have come to realize that in many important mathematical systems there exist operations with analogous properties and, furthermore, that the mathematical theories of these systems are essentially dependent on these laws of combination. It is far beyond the scope of this book to discuss these theories; it must suffice to say that they occur in the extension of number theory to other systems than the ordinary integers; they appear in many theories of algebra, in function theory and geometry, and even in logic. In all cases, there are two operations corresponding in our special case to g.c.d. and l.c.m. A special notation has been introduced for such operations, namely,

$$d = a \cap b, \qquad m = a \cup b$$

while various names are in use, for instance, *meet* and *join* or *union* and *cross-cut*. The two operations satisfy the same dual set of axioms as those mentioned previously:

1.

$$a \cap a = a, \qquad\qquad a \cup a = a$$

2.

$$a \cap b = b \cap a, \qquad\qquad a \cup b = b \cup a$$

3.

$$a \cap (b \cap c) = (a \cap b) \cap c, \qquad a \cup (b \cup c) = (a \cup b) \cup c$$

4.

$$a \cup (a \frown b) = a, \qquad\qquad a \cap (a \cup b) = a$$

Systems that satisfy these axioms have been called *lattices* or sometimes *structures*.

Besides the g.c.d. and the l.c.m. in number theory we shall mention only a single other example of such systems. Let A and

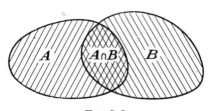

Fig. 5–2.

B be two *sets* of points or, more general, of elements of some sort. One can picture A and B as the shaded portions of the plane in the illustration (see Fig. 5–2). Then the *union* or *sum* $A \cup B$ of the two sets consists of the elements that belong to either A or B, while the *cross-cut* or *intersection* $A \cap B$ contains the common elements to A and B. In the figure $A \cup B$ is the whole shaded part of the plane while $A \cap B$ is doubly shaded. The reader will have no difficulty in verifying that the four pairs of axioms for a lattice are satisfied.

In most such lattice systems there are further rules which the two operations satisfy. This is, for instance, the case in the theory we are particularly interested in here, namely, the g.c.d. and l.c.m. of numbers. The two subsequent theorems can be interpreted as giving laws of this kind.

THEOREM 5–9. For any three integers a, b, and c one has the rules

$$(a, [b, c]) = [(a, b), (a, c)] \qquad (5\text{–}33)$$

and

$$[a, (b, c)] = ([a, b], [a, c]) \qquad (5\text{–}34)$$

connecting the greatest common divisor and the least common multiple. One is the dual of the other.

One can express the law (5–33) in the form that the g.c.d. of a number a with the l.c.m. of two numbers b and c may be found by computing the g.c.d. of a with b and c separately and taking the l.c.m. of the results. The second rule (5–34) can be stated analogously. These laws are commonly called the *distributive laws*.

To prove the equality (5–33) it is probably simplest to use the unique factorization theorem and verify for each prime that the exponents to the various powers to which it is raised on both sides are the same. We assume that some prime p divides the numbers a, b, and c to powers with the exponents α, β, and γ, respectively. Since b and c appear symmetrically in (5–33), there is no limitation in arranging the notation such that $\beta \geq \gamma$. Then the exponent of the power of p contained in $[b, c]$ is β. Consequently, the left-hand side of (5–33) contains p to a power with the exponent

$$\min (\alpha, \beta) \qquad (5\text{–}35)$$

On the other hand, in (a, b) and (a, c) the prime p occurs with the exponents

$$\min (\alpha, \beta), \qquad \min (\alpha, \gamma)$$

Since $\beta \geq \gamma$ the first of these numbers is the larger, so that the right-hand side of (5–33) also contains p to a power with the exponent (5–35). This completes the proof of (5–33), and the equality (5–34) may be derived quite analogously.

There exists another interesting identity, which we shall now derive:

THEOREM 5–10. For any three integers a, b, and c one has

$$([a, b], [a, c], [b, c]) = [(a, b), (a, c), (b, c)] \qquad (5\text{–}36)$$

This formula is peculiar in that it states that a certain expression involving the g.c.d. and l.c.m. of three numbers is *self-dual, i.e.,* it remains the same when the two operations are interchanged. As before, to prove (5–36), let α, β, and γ be the exponents of the powers to which some prime p divides a, b, and c. Since the expression (5–36) is symmetrical in a, b, and c, we can arrange the notation such that

$$\alpha \geqq \beta \geqq \gamma$$

Then $[a, b]$, $[a, c]$, and $[b, c]$ contain p to powers with the exponents α, α, and β, respectively. In their g.c.d. the exponent therefore is β. On the other hand, (a, b), (a, c), and (b, c) have the exponents β, γ, and γ, respectively, for p so that their l.c.m. contains p to a power with the exponent β. Thus for any prime p both sides in (5–36) contain the same power and consequently they are equal. One could also have derived (5–36) by using the rules in theorem 5–9.

It should be noted that the rules (5–33) and (5–34) can be considered to be equivalent to properties of maxima and minima of three numbers α, β, and γ, namely:

$$\min \{\alpha, \max (\beta, \gamma)\} = \max \{\min (\alpha, \beta), \min (\alpha, \gamma)\}$$

$$\max \{\alpha, \min (\beta, \gamma)\} = \min \{\max (\alpha, \beta), \max (\alpha, \gamma)\}$$

Similarly the result in theorem 5–10 is a consequence of

$$\min \{\max (\alpha, \beta), \max (\alpha, \gamma), \max (\beta, \gamma)\}$$
$$= \max \{\min (\alpha, \beta), \min (\alpha, \gamma), \min (\beta, \gamma)\}$$

a law that remains invariant when maximum and minimum are interchanged.

It is worth noting that the rules we have obtained in theorems 5–9 and 5–10 for the g.c.d. and l.c.m. have a much greater generality than may appear from their derivation in our special field. These relations form a part of many other mathematical theories of importance.   We shall mention only one instance, namely, the case of *set operations*, which we introduced above.   Let us recall the notations that when A and B are sets, $A \cup B$ denotes the union or sum set while $A \cap B$ is their intersection or common part. Now let A, B, and C be three arbi-

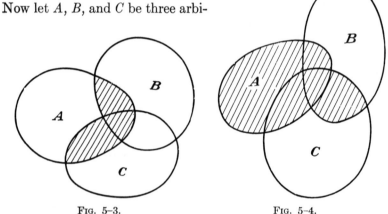

Fig. 5–3.   Fig. 5–4.

trary sets.   We shall see how the relations become evident on the basis of simple illustrations.

In our notation the distributive law (5–33) takes the form

$$A \cap (B \cup C) = (A \cap B) \cup (A \cap C)$$

In Fig. 5–3 the shaded part is the set of points common to A and the sum of B and C.   Clearly this set may also be considered the sum of two sets, namely, the common part of A and B and the common part of A and C.

Corresponding to (5–34) one obtains

$$A \cup (B \cap C) = (A \cup B) \cap (A \cup C)$$

In Fig. 5–4 the sum of A and the common part of B and C has been shaded.   But it is evident that this set is also the common part of the sum sets of A and B and of A and C, as the formula requires.

The analogue of the relation (5–36) is

$$(A \cup B) \cap (A \cup C) \cap (B \cup C)$$
$$= (A \cap B) \cup (A \cap C) \cup (B \cap C)$$

In Fig. 5–5 the shaded part is found by inspection to consist of all points that are common to the three sum sets

$$A \cup B, \quad A \cup C, \quad B \cup C$$

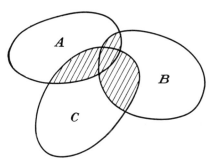

FIG. 5–5.

But it is also obvious that it is the sum of three sets, namely, the intersection sets

$$A \cap B, \quad A \cap C, \quad B \cap C$$

as we wanted to verify.

Problems.

1. Verify the relations in theorems 5–9 and 5–10 for the three numbers: 60, 72, 96.

2. The relations in the two theorems involve only the g.c.d. and the l.c.m There exist several other relations that also contain multiplication and division The reader may attempt to verify the relatively complicated identity

$$(ab,\ cd) = (a,\ c)(b,\ d)\left(\frac{a}{(a,\ c)},\ \frac{d}{(b,\ d)}\right)\left(\frac{c}{(a,\ c)},\ \frac{b}{(b,\ d)}\right)$$

5–5. Euler's function. When m is some integer, we shall consider the problem of finding how many of the numbers

$$1, 2, 3, \ldots, m - 1, m \tag{5–37}$$

are relatively prime to m. This number is usually denoted by $\varphi\,(m)$, and it is known as *Euler's φ-function* of m because Euler around 1760 for the first time proposed the question and gave its solution. Other names, for instance, *indicator* or *totient* have occasionally been used.

Example.

Among the positive integers less than 42 there are 12 that are relatively prime to 42, namely:

$$1,\, 5,\, 11,\, 13,\, 17,\, 19,\, 23,\, 25,\, 29,\, 31,\, 37,\, 41$$

so that $\varphi(42) = 12$.

For the first few integers one finds

$$\varphi(2) = 1, \qquad \varphi(6) = 2, \qquad \varphi(10) = 4$$
$$\varphi(3) = 2, \qquad \varphi(7) = 6, \qquad \varphi(11) = 10$$
$$\varphi(4) = 2, \qquad \varphi(8) = 4, \qquad \varphi(12) = 4$$
$$\varphi(5) = 4, \qquad \varphi(9) = 6,$$

The value $\varphi(1)$ is in itself without meaning, but by special definition one puts $\varphi(1) = 1$.

In some cases the determination of $\varphi(m)$ is particularly simple. When $m = p$ is a prime, all numbers (5–37) except the last are relatively prime to p; consequently,

$$\varphi(p) = p - 1$$

When $m = p^\alpha$ is a power of a prime, the only numbers in (5–37) that have a common factor with m are the multiples of p,

$$p,\, 2p,\, \ldots,\, p^{\alpha-1}p$$

Since there are $p^{\alpha-1}$ of these multiples, one has

$$\varphi(p^\alpha) = p^\alpha - p^{\alpha-1} = p^{\alpha-1}(p - 1) = p^\alpha\left(1 - \frac{1}{p}\right) \quad (5\text{–}38)$$

We shall now tackle the general case. Let p be some prime dividing m and let us first find the number $\varphi_p(m)$ of integers in

(5–37) that are not divisible by p. This is simple since those that
are divisible by p are the multiples

$$p, 2p, \ldots, \frac{m}{p} \cdot p \qquad (5\text{–}39)$$

Since there are m/p of them, the remaining ones not divisible by
p are, in number,

$$\varphi_p(m) = m - \frac{m}{p} = m\left(1 - \frac{1}{p}\right) \qquad (5\text{–}40)$$

Next let q be some other prime dividing m. To find the number
$\varphi_{pq}(m)$ of integers in (5–37) divisible neither by p nor q, one must
deduct from $\varphi_p(m)$ the m/q multiples of q

$$q, 2q, \ldots, \frac{m}{q} q \qquad (5\text{–}41)$$

But this procedure involves a duplication since the multiples
(5–39) and (5–41) have some elements in common, namely, the
m/pq multiples of pq,

$$pq, 2pq, \ldots, \frac{m}{pq} pq$$

Instead of deducting m/q, we must only subtract

$$\frac{m}{q} - \frac{m}{pq} = \frac{m}{q}\left(1 - \frac{1}{p}\right)$$

from $\varphi_p(m)$. Therefore the total of integers in (5–37) not divisible
by p or q is

$$\varphi_{p,q}(m) = m\left(1 - \frac{1}{p}\right) - \frac{m}{q}\left(1 - \frac{1}{p}\right) = m\left(1 - \frac{1}{p}\right)\left(1 - \frac{1}{q}\right)$$

$$(5\text{–}42)$$

We take the final step by mathematical induction. The results
in (5–40) and (5–42) show that for $t = 1$ and $t = 2$, if $p_1, p_2, \ldots, p_t$
are primes dividing m, there are

$$\varphi_{p_1,\ldots,p_t}(m) = m\left(1 - \frac{1}{p_1}\right)\left(1 - \frac{1}{p_2}\right)\cdots\left(1 - \frac{1}{p_t}\right) \qquad (5\text{–}43)$$

integers in (5–37) that are divisible by none of them. We assume that this formula (5–43) has been established for t primes and wish to show that it is correct when there is one further prime p_{t+1}. To determine the number of integers in (5–37) that are not divisible by any of the $p_1, \ldots, p_t$ nor by p_{t+1}, one must subtract from (5–43) the number of multiples of p_{t+1}

$$p_{t+1}, 2p_{t+1}, \ldots, \frac{m}{p_{t+1}} \cdot p_{t+1} \tag{5–44}$$

which have not already been stricken out. A multiple

$$a p_{t+1}, \qquad a = 1, 2, \ldots, \frac{m}{p_{t+1}}$$

has, however, been taken into consideration previously if, and only if, a is divisible by one of the primes $p_1, \ldots, p_t$. But according to our formula (5–43), there are

$$\frac{m}{p_{t+1}}\left(1 - \frac{1}{p_1}\right) \cdots \left(1 - \frac{1}{p_t}\right)$$

numbers in the series

$$1, 2, \ldots, \frac{m}{p_{t+1}}$$

not divisible by any $p_1, \ldots, p_t$. One concludes that in (5–37) there are

$$\varphi_{p_1,\ldots,p_{t+1}}(m) = m\left(1 - \frac{1}{p_1}\right) \cdots \left(1 - \frac{1}{p_t}\right)$$
$$- \frac{m}{p_{t+1}}\left(1 - \frac{1}{p_1}\right) \cdots \left(1 - \frac{1}{p_t}\right)$$
$$= m\left(1 - \frac{1}{p_1}\right) \cdots \left(1 - \frac{1}{p_t}\right)\left(1 - \frac{1}{p_{t+1}}\right)$$

numbers not divisible by any $p_1, p_2, \ldots, p_{t+1}$. This establishes our rule (5–43) in general.

To obtain Euler's φ-function let

$$m = p_1^{\alpha_1} \cdots p_r^{\alpha_r} \tag{5–45}$$

be the prime factorization of the given number. An integer in (5–37) is relatively prime to m only when it is not divisible by any of the primes $p_1, \ldots, p_r$. The result that we just derived yields, therefore:

THEOREM 5–11. Let m be an integer whose various prime factors are $p_1, \ldots, p_r$. Then there are

$$\varphi(m) = m \left(1 - \frac{1}{p_1}\right) \cdots \left(1 - \frac{1}{p_r}\right) \tag{5–46}$$

integers less than and relatively prime to m.

Example.

For $m = 42$ the prime factors are 2, 3, and 7, so that

$$\varphi(42) = 42 \cdot (1 - \tfrac{1}{2}) \cdot (1 - \tfrac{1}{3}) \cdot (1 - \tfrac{1}{7}) = 12.$$

A table of the values of the φ-function for all numbers up to 10,000 has been computed by J. W. L. Glaisher.

The expression (5–46) for Euler's φ-function can be written in slightly different forms. By means of the prime factorization (5–45), one finds

$$\begin{aligned}
\varphi(m) &= p_1{}^{\alpha_1} \cdots p_r{}^{\alpha_r} \left(1 - \frac{1}{p_1}\right) \cdots \left(1 - \frac{1}{p_r}\right) \\
&= p_1{}^{\alpha_1-1}(p_1 - 1) \cdots p_r{}^{\alpha_r-1}(p_r - 1) \\
&= (p_1{}^{\alpha_1} - p_1{}^{\alpha_1-1}) \cdots (p_r{}^{\alpha_r} - p_r{}^{\alpha_r-1})
\end{aligned}$$

When the last expression is compared with the formula (5–38) for the φ-function of a prime power, one sees that

$$\varphi(m) = \varphi(p_1{}^{\alpha_1}) \cdots \varphi(p_r{}^{\alpha_r})$$

From this result one concludes further that when

$$m = a \cdot b$$

where a and b are relatively prime, one has

$$\varphi(m) = \varphi(a) \cdot \varphi(b)$$

This is analogous to the property expressed in (5-8) for the sum $\sigma(m)$ of the divisors of a number.

We shall deduce only one further fact about Euler's function:

THEOREM 5–12. When d runs through all divisors of a number m, the sum of all the corresponding φ-function values $\varphi(d)$ is equal to m

$$\sum\varphi(d) = m \qquad (5\text{-}47)$$

Before we establish a proof of this theorem we shall illustrate it.

Example.

The number

$$m = 42 = 2 \cdot 3 \cdot 7$$

has eight divisors, namely,

$$d = 1, 2, 3, 6, 7, 14, 21, 42$$

For these one finds

$$\varphi(1) = 1, \qquad \varphi(7) = 6$$
$$\varphi(2) = 1, \qquad \varphi(14) = 6$$
$$\varphi(3) = 2, \qquad \varphi(21) = 12$$
$$\varphi(6) = 2, \qquad \varphi(42) = 12$$

and the sum of these values is $m = 42$.

In the special case where $m = p^\alpha$ is a power of a prime, the proof of the relation (5-47) is particularly simple. The divisors of m are

$$1, p, p^2, \ldots, p^\alpha$$

and one has

$$\varphi(1) + \varphi(p) + \varphi(p^2) + \varphi(p^3) + \cdots + \varphi(p^\alpha)$$
$$= 1 + (p-1) + (p^2-p) + (p^3-p^2) + \cdots + p^\alpha - p^{\alpha-1} = p^\alpha \quad (5\text{-}48)$$

since all terms except p^α cancel out.

This result (5–48) facilitates the proof of the general theorem. We assume that m has the prime factorization (5–45) and form the product of the following expressions:

$$[\varphi(1) + \varphi(p_1) + \varphi(p_1^2) + \cdots + \varphi(p_1^{\alpha_1})]$$
$$\times [\varphi(1) + \varphi(p_2) + \varphi(p_2^2) + \cdots + \varphi(p_2^{\alpha_2})]$$
$$\cdot \quad \cdot \quad \cdot \quad \cdot \quad \cdot \quad \cdot \quad \cdot \quad \cdot \quad \cdot \quad \cdot \quad \cdot \quad \cdot \quad \cdot \quad \cdot \quad \cdot \quad \cdot \quad \cdot$$
$$\times [\varphi(1) + \varphi(p_r) + \varphi(p_r^2) + \cdots + \varphi(p_r^{\alpha_r})]$$

According to (5–48) these brackets are respectively equal to

$$p_1^{\alpha_1}, p_2^{\alpha_2}, \ldots, p_r^{\alpha_r}$$

so that their product is m as in the right-hand side of (5–47). When the product of the brackets is formed, one takes one term from each of them and obtains all expressions

$$\varphi(p_1^{\delta_1})\varphi(p_2^{\delta_2}) \cdots \varphi(p_r^{\delta_r}) = \varphi(p_1^{\delta_1}p_2^{\delta_2} \cdots p_r^{\delta_r})$$

where each δ_i may take some value 0, 1, ..., α_i. But this means that the numbers

$$d = p_1^{\delta_1}p_2^{\delta_2} \ldots p_r^{\delta_r}$$

run through all divisors of m. The sum of all the terms resulting from the multiplication of the brackets is therefore $\sum\varphi(d)$, and the formula (5–47) has been established.

Problems.

1. Determine $\varphi(N)$ and verify theorem 5–12 for the numbers (a) $N = 120$, (b) $N = 365$.

2. Prove the formula

$$\varphi(m^2) = m \cdot \varphi(m)$$

3. Find all numbers m such that $\varphi(m)$ divides m.

Bibliography

Escott, E. B.: "Amicable Numbers," *Scripta Mathematica*, Vol. 12, 61–72 (1946).

Glaisher, J. W. L.: *Number-divisor Tables*, Vol. 8, British Association Mathematical Tables, Cambridge, 1940.

Poulet, P.: *La Chasse aux nombres*, 2 vols. Brussels, 1929, 1934.

CHAPTER 6

INDETERMINATE PROBLEMS

6–1. Problems and puzzles. Riddles, puzzles, and trick questions constitute a part of the folklore over large parts of the world. Curiously enough, the American Indians seem to have had no feeling for this form of entertainment, since no trace of such problem lore has been found among them by anthropologists. However, in Europe, Africa, and Asia one finds a multitude of such problems; they spread with the cultural interchanges, often with a preservation of details that is remarkable. Many of the puzzles have a germ of mathematical content. They can often be recognized as interrelated by the faithfulness with which certain figures and forms of questions are reproduced, even in localities separated widely in time and place.

Let us illustrate these observations by an outstanding example. One of our most important sources of ancient Egyptian mathematical knowledge is the *Papyrus Rhind*, now in the British museum. It is usually named for its previous owner, the Egyptologist Henry Rhind, or sometimes for the scribe Ahmes who copied it from earlier sources about 1800 B.C. The letters and figures are in hieratic writing. The papyrus contains in somewhat systematic arrangement the solution of a variety of problems, many of them practical everyday questions not very different from those still encountered in our present-day school texts. But the almost magical esteem of early mathematical learning may be seen in the introductory statement to the manuscript wherein it is promised to give "directions for obtaining knowledge of all obscure things."

Most computations in the *Papyrus Rhind* have been relatively easy to decipher. However towards the end one finds a curious problem where the interpretation is not so certain. When transcribed it consists of a column of terms (see Fig. 6–1).

Houses	7
Cats	49
Mice	343
Ears of wheat	2,401
Hekat measure	16,807
Total	19,607

It is preceded by a word that seems to mean estate, and in a secondary column the same answer, 19,607, has been obtained in a different manner. The figures one recognizes, of course, as the five powers of the number 7. Some commentators assumed directly that this would show that the Egyptian mathematicians had mastered the concept of powers at this early time and that the names houses, cats, and so on, were symbolic terms for powers of the various orders.

However, scholars more familiar with the history of mathematics have pointed out that problems with the same figures are known from other sources. We have already mentioned the importance of the *Liber abaci* (A.D. 1202) by Leonardo Pisano in the introduction of Arabic numerals to Europe. Among the problems included in this work is the following curious one: Seven old women on the road to Rome, each woman has seven mules, each mule carries seven sacks, each sack contains seven loaves, with each loaf there are seven knives, and each knife is in seven sheaths. How many objects are there, women, mules, sacks, loaves, knives, and sheaths? Clearly Leonardo has borrowed this problem from the medieval entertainment lore of his time, 3,000 years after the compilation of the reckoning book of Ahmes. But one need not go so far abroad. The old English children's rhyme contains a jocular problem again based on the powers of the sacred number 7

"As I was going to St. Ives,
I met a man with seven wives
Every wife had seven sacks
Every sack had seven cats
Every cat had seven kits
Kits, cats, sacks, and wives
How many were going to St. Ives?"

According to the first line none of them was going to St. Ives as one would gleefully reveal to the victim after he had performed the lengthy computations. It seems likely that Leonardo's problem must have had the same twist of surprise in its original popular version.

This brings us back to Ahmes's old problem, and it appears quite likely that it may have been introduced, like Leonardo's, because the figures were familiar from a trick question. As to a suitable formulation the reader may use his own imagination; perhaps it might have run somewhat as follows: "A man's estate included seven houses, each house had seven cats, for each cat there were seven mice, for each mouse there were seven ears of wheat, and each ear would yield seven measures of grain. How many things did he possess, houses, cats, mice, ears, and measures all?" Perhaps the answer finally should be none, because the owner was dead.

Numerous other examples of the preservation of the ideas of mathematical puzzles may be given. Several medieval manuscripts containing collections of popular problems are still extant, and many of these puzzles, with small variations, may be recognized in our present-day magazines almost every week. Let us mention only a few.

1. An old woman goes to market and a horse steps on her basket and crushes the eggs. The rider offers to pay for the damages and asks her how many eggs she had brought. She does not remember the exact number, but when she had taken them out two at a time, there was one egg left. The same happened when she picked them out three, four, five, and six at a time, but

FIG. 6–1. Problem from *Papyrus Rhind.*
Above, hieratic original; *below,* hieroglyphic transcription. Note that writing is from right to left. (*Courtesy of Mathematical Association of America.*)

when she took them out seven at a time they came out even. What is the smallest number of eggs she could have had?

2. A cistern can be filled by one pipe in one hour, by another in two, by a third in three, and by a fourth in four. How long a time will it take for all four pipes together to fill it?

3. Two men have a full eight-gallon jug of wine and also two empty jugs taking five and three gallons. How can they divide the wine evenly?

4. Three jealous husbands must ferry a river with their wives. There is only one small skiff capable of taking two persons at a time. How can one transport all six across so that no wife is ever left with other men without her husband's being present?

5. Fifteen Christians and fifteen Turks were passengers on the same ship when a terrible storm arose. To lighten the ship, the captain orders that half the passengers should be thrown overboard to save the others. All 30 are placed in a circle, and one agrees to count out every ninth person and throw him overboard. Providence intervenes, and it turns out that all the Turks are thrown overboard and the Christians saved. How had the passengers been arranged?

The importance of such problems as a matter of diversion has of course decreased, and certainly they cannot compete with our modern mechanized amusement industry; but as a part of our folklore they are far from extinct. An interesting side light on this fact came during the last war when long waiting was the most nerve-racking and the most common activity of the soldiers. At this time, as we already mentioned, teachers of mathematics received a surprising number of requests from servicemen both for the correct answers and for methods for solving puzzle problems. Often they had arrived independently at the solution through the most laborious guesses.

6–2. Indeterminate problems. There is a type of problem that occurs quite commonly in puzzles and whose theory constitutes a particularly significant part of number theory. These problems may appropriately be called *linear indeterminate problems*, for reasons that will become clear after some examples.

One of the earliest occurrences of such problems in Europe is to be found in a manuscript containing mathematical problems dating from about the tenth century. It is believed possible that it may be a copy of a collection of puzzles which Alcuin prepared for Charlemagne. The problem we are interested in runs as follows:

1. When 100 bushels of grain are distributed among 100 persons so that each man receives three bushels, each woman two bushels, and each child half a bushel, how many men, women, and children are there?

To formulate this problem mathematically let x, y, and z denote the number of men, women, and children, respectively. The conditions of the problem then give

$$x + y + z = 100, \qquad 3x + 2y + \tfrac{1}{2}z = 100 \qquad (6\text{--}1)$$

As we shall see later there are several solutions but Alcuin gives only the values

$$x = 11, \qquad y = 15, \qquad z = 74$$

From an Arabic manuscript copied about A.D. 1200, but undoubtedly composed earlier, we take this example:

2. One duck may be bought for 5 drachmas, one chicken for 1 drachma, and 20 starlings for 1 drachma. You are given 100 drachmas and ordered to buy 100 birds. How many will there be of each kind?

When x, y, and z are the number of ducks, chickens, and starlings, it follows that

$$x + y + z = 100, \qquad 5x + y + \frac{z}{20} = 100 \qquad (6\text{--}2)$$

One may observe that the same number occurs on the right-hand side in both equations (6–1) and in (6–2). This particular preference in the choice of the figures in the questions is common in Arabic, Chinese, and medieval European problems, and it undoubtedly points to an interrelated or common background.

Even the use of the special number 100 shows a peculiar persistence in problems from all these sources.

One finds similar questions in the many medieval collections of problems. They occur in Leonardo's *Liber abaci* (A.D. 1202), probably derived from Arabic sources, and in the following centuries they became increasingly popular. To illustrate a fairly common type of formulation we quote from a German reckoning manual (Christoff Rudolff, 1526):

3. At an inn, a party of 20 persons pay a bill for 20 groschen. The party consists of men (x), women (y), and maidens (z), each man paying 3, each woman 2, and each maiden $\frac{1}{2}$ groschen. How was the party composed?

Here the equations become

$$x + y + z = 20, \qquad 3x + 2y + \frac{z}{2} = 20 \qquad (6\text{–}3)$$

and the figures are so chosen that there is a unique solution $x = 1$, $y = 5$, $z = 14$.

It is, of course, not certain that this type of problem originated within a single cultural sphere, but if so, it seems likely that India should be looked to for its source. As early as the arithmetic of Aryabhata (around A.D. 500) one finds indeterminate problems. Brahmagupta (born A.D. 598) in his mathematical and astronomical manual *Brahma-Sphuta-Siddhanta* ("Brahma's correct system") not only introduces them, but gives a perfected method for their solution that is practically equivalent to our present procedures. The method is called the *cuttaca* or *pulverizer* and is based upon Euclid's algorism. Brahmagupta's examples are almost all of astronomical character and refer to the comparisons between periods of revolution of the heavenly bodies and determinations of their relative positions.

We take the following problem from the *Lilavati* by Bhaskara, a work we have already mentioned:

4. Say quickly, mathematician, what is the multiplier by which two hundred and twenty-one being multiplied and sixty-five added

to the product the sum divided by one hundred and ninety-five becomes exhausted?

Here one wishes to find some x satisfying the condition

$$221x + 65 = 195y \qquad (6\text{-}4)$$

In the *Lilavati*, as well as in other Hindu treatises on mathematics, one finds many problems in the flowery style so customary in Hindu writings. This problem is from the *Bija-Ganita*, literally meaning *seed-counting* but denoting *algebra*, also composed by Bhaskara:

5. The quantity of rubies without flaw, sapphires, and pearls belonging to one person is five, eight, and seven respectively; the number of like gems appertaining to another is seven, nine and six; in addition, one has ninety-two coins, the other sixty-two and they are equally rich. Tell me quickly then, intelligent friend, who art conversant with algebra, the prices of each sort of gem.

In Hindu mathematics colors were used to denote the various unknowns, black, blue, yellow, red, and so on. If we prosaically denote the prices of rubies, sapphires, and pearls by x, y, and z, the condition becomes

$$5x + 8y + 7z + 92 = 7x + 9y + 6z + 62 \qquad (6\text{-}5)$$

A further example may be taken from Mahaviracarya's work *Ganita-Sara-Sangraha*, probably composed around A.D. 850:

6. Into the bright and refreshing outskirts of a forest which were full of numerous trees with their branches bent down with the weight of flowers and fruits, trees such as jambu trees, date-palms, hintala trees, palmyras, punnaga trees and mango trees—filled with the many sounds of crowds of parrots and cuckoos found near springs containing lotuses with bees roaming around them—a number of travelers entered with joy.

There were 63 equal heaps of plantain fruits put together and seven single fruits. These were divided evenly among 23 travelers. Tell me now the number of fruits in each heap.

It is quite an anticlimax to state that if x is the number of fruits in each heap, one must have

$$63x + 7 = 23y \qquad (6\text{--}6)$$

This beautiful Hindu forest contains a number of other problems, but after all these ancient examples let us conclude with one with a more modern touch. The following letter was one among several similar ones received by the author during the recent war:

Dear Sir:

A group of bewildered GI's at Guadalcanal, most of whom have been out of school for a good many years and have forgotten how to solve algebraic problems, have been baffled by what appears to be a very simple problem. Some of them affirm that it cannot be worked other than through the trial and error method, but I maintain that it can be worked systematically by means of some sort of formula or equation.

[7.]   Here is the problem: A man has a theater with a seating capacity of 100. He wishes to admit 100 people in such a proportion that will enable him to take in \$1.00 with prices as follows: men 5¢, women 2¢, children 10 for one cent. How many of each must be admitted?

Can this problem be solved other than through the laborious trial and error method? We shall greatly appreciate your assistance in helping us to find the solution, thus relieving our weary brains.

Yours truly,

. . . .

P.S.   Through the trial and error method we found the answer to be 11 men, 19 women, and 70 children.

In terms of equations we have the conditions

$$x + y + z = 100, \qquad 5x + 2y + \frac{z}{10} = 100 \qquad (6\text{--}7)$$

Here again our familiar number 100 figures on the right. It is also clear that the figures are not adapted to the present movie prices and that we are confronted with an ancient problem which has gained in actuality by being put in modern dress.

6–3. Problems with two unknowns. We have presented a whole series of examples of linear indeterminate problems. As we saw, they lead to one or more linear equations between the unknown quantities. Furthermore, the number of unknowns is greater than the number of equations so that if there were no limitations on the kind of values the solutions could take, one could give arbitrary values to some of the unknowns and find the others in terms of them. For instance, in problem 4 one could simply write

$$x = \frac{195y - 65}{221}$$

and any value of y would give a corresponding value of x. However, by the terms of the problems, the choice of solutions is limited to integral values and usually also to positive numbers. But even with these restrictions, the solutions may be indeterminate in the sense that there may be several, or even an infinite number of them, as we shall see. On the other hand there may be no solution at all.

Clearly many of our previous problems could be solved by probing, by trial and error, and in medieval times this procedure must have been commonly used. In several problems the possibilities are rather limited so that not many attempts need to be made. We have already mentioned that a method for solving linear indeterminate problems was found quite early by the Hindu school of mathematics. In Europe a corresponding method was not discovered until a millenium later, and the date of rediscovery can be fixed quite accurately. In 1612 there appeared in Lyons a collection of ancient puzzles under the title: *Problèmes plaisans et delectables, qui se font par les nombres.* The author was Claude-Gaspar Bachet, Sieur de Méziriac (1581–1638), a gentleman, scholar, poet, and theologian, ardently devoted to classical learning. His work proved popular and a second enlarged edition appeared in 1624. Here one finds for the first time his rules for solving indeterminate problems.

One of Bachet's problems runs about as follows:

8. A party of 41 persons, men, women, and children, take part in a meal at an inn. The bill is for 40 sous and each man pays 4

sous, each woman 3, and every child $\frac{1}{3}$ sou. How many men, women, and children were there?

In this case we have the equations

$$x + y + z = 41, \qquad 4x + 3y + \frac{z}{3} = 40 \qquad (6\text{--}8)$$

Bachet's procedure unfortunately is complicated by a lack of algebraic symbolism. In the following sections we shall make a more systematic study of the linear indeterminate problems. Here we prefer to present a method of repeated reductions that is easy to explain. It works quite well when the numbers involved are not too large, as, for instance, in most of the examples we have already mentioned. This method was used extensively by Euler in his popular *Algebra* (1770), which devotes much space to indeterminate problems.

We shall deal first with a single linear equation

$$ax + by = c \qquad (6\text{--}9)$$

in two unknowns. As a preliminary example we take simply

$$x + 7y = 31 \qquad (6\text{--}10)$$

which may be written

$$x = 31 - 7y \qquad (6\text{--}11)$$

This shows that any integral value substituted for y in (6–11) will give an integral value for x; for instance, $y = 6, x = -11$; or $y = 0$, $x = 31$. Thus there will be an infinite set of pairs of solutions. But if one requires positive solutions, one must have both $y > 0$ and

$$x = 31 - 7y > 0$$

thus $y < 4\frac{3}{7}$. This gives only four possibilities, $y = 1, 2, 3, 4$, with the corresponding values, $x = 24, 17, 10, 3$, for the other unknown.

This trivial example was introduced in order to show that when one of the coefficients of x and y in (6–9) is unity, as in (6–10), the solution is immediate. The guiding principle in the method used below is to reduce the more general equations in successive steps to this simple form. The first example given by Euler is:

Write the number 25 as the sum of two (positive) integers, one divisible by 2 and the other by 3.

The two summands may be taken to be $2x$ and $3y$ so that

$$2x + 3y = 25 \qquad (6\text{-}12)$$

is the equation to be fulfilled. Since x has the smaller coefficient, we solve for x and find by taking out the integral parts of the fractional coefficients

$$x = \frac{25 - 3y}{2} = 12 - 2y + \frac{1 + y}{2} \qquad (6\text{-}13)$$

Because x and y are integers, the quotient

$$t = \frac{1 + y}{2} \qquad (6\text{-}14)$$

is integral. Conversely, any integral value t we may give to this quotient (6-14) will make y integral

$$y = 2t - 1$$

and also x integral according to (6-13)

$$x = 12 - 2y + t = 14 - 3t$$

This shows that the general integral solution of (6-12) is

$$x = 14 - 3t, \qquad y = 2t - 1$$

and one can verify by substitution that they actually satisfy the equation. Consequently, there is an infinite number of solutions, one for each integral t. For instance, when $t = 10$, $x = -16$, $y = 19$.

But if one is limited to positive values, one must have

$$x = 14 - 3t > 0, \qquad y = 2t - 1 > 0$$

hence

$$\tfrac{1}{2} < t < 4\tfrac{2}{3}$$

and there are only four permissible values, $t = 1, 2, 3, 4$. The corresponding solutions are

$$x = 11, 8, 5, 2$$

$$y = \ \ 1, 3, 5, 7$$

This gives the decompositions

$$25 = 22 + 3 = 16 + 9 = 10 + 15 = 4 + 21$$

required in the original problem, as one could have verified without much effort by probing.

This example requires only one reduction. In most cases two or more steps are required. Let us illustrate this by another example taken from Euler's *Algebra*.

A man buys horses and cows for a total amount of \$1,770. One horse costs \$31 and one cow \$21. How many horses and cows did he buy?

When x is the number of horses and y the number of cows, the condition

$$31x + 21y = 1{,}770 \tag{6–15}$$

must be fulfilled. Here y has the smaller coefficient so we solve for y and find

$$y = \frac{1{,}770 - 31x}{21} = 84 - x + \frac{6 - 10x}{21} \tag{6–16}$$

This requires that the quotient

$$t = \frac{6 - 10x}{21}$$

shall be integral. Our task is, therefore, to find integers x and t such that

$$21t + 10x = 6 \tag{6–17}$$

As can be seen, this is an equation of the same type as (6–15) but with smaller numbers so that a first reduction has been performed. Since x has the smaller coefficient, we derive from (6–17)

$$x = -2t + \frac{6 - t}{10} \qquad (6\text{–}18)$$

We conclude that x can only be integral when

$$u = \frac{6 - t}{10}$$

is integral or

$$t = 6 - 10u$$

for some integer u. By substituting this value into (6–18) and then x into (6–16), one finds

$$x = -12 + 21u, \qquad y = 102 - 31u$$

Any integral value of u will give integers x and y satisfying the equation (6–15) so that we have obtained the general solution. The form of the problem requires, however, that x and y must be positive. This leads to the conditions

$$-12 + 21u > 0, \qquad 102 - 31u > 0$$

or

$$\tfrac{12}{21} < u < 3\tfrac{9}{31}$$

There are, therefore, three possible values $u = 1, 2, 3$, and the corresponding solutions are

$$x = 9, 30, 51$$

$$y = 71, 40, 9$$

We could have made the solution of the problem unique, for instance, by requiring in the formulation that the number of horses would be greater than the number of cows.

As a last example of this type we shall take problem 4 in the preceding section, stated by Bhaskara. It is of interest because it permits us to mention some simplifications that often are available in the solution of indeterminate problems.

We observe first that in (6–4) the coefficients 221, 65, and 195 are all divisible by 13. This factor can, therefore, be canceled and the equation becomes

$$17x + 5 = 15y \qquad (6\text{--}19)$$

Here, furthermore, both 5 and $15y$ are divisible by 5 so that $17x$ must have this factor. But 17 is prime to 5 so that x must be divisible by 5, and we can write

$$x = 5x_1$$

When this is substituted in (6–19), one can cancel by 5 and have the still simpler equation

$$17x_1 + 1 = 3y$$

By writing this

$$y = \frac{17x_1 + 1}{3} = 6x_1 + \frac{1 - x_1}{3}$$

we see that

$$\frac{1 - x_1}{3} = t$$

is integral. This gives $x_1 = 1 - 3t$ and

$$x = 5x_1 = 5 - 15t, \qquad y = 6 - 17t \qquad (6\text{--}20)$$

as the general solution.

Let us ask for the positive solutions. One obtains, as previously, the conditions

$$5 - 15t > 0, \qquad 6 - 17t > 0$$

or

$$t < \tfrac{1}{3}, \qquad t < \tfrac{6}{17}$$

This shows that all values $t = 0, -1, -2, \cdots$ will give positive solutions in (6–20). To obtain positive values for this parameter or auxiliary variable, it is convenient in (6–20) to write $t = -u$ so that

$$x = 5 + 15u, \qquad y = 6 + 17u$$

becomes the general solution; all values $u = 0, 1, 2, \cdots$ give positive answers, namely,

$$x = 5, 20, 35, 50, \cdots$$

$$y = 6, 23, 40, 57, \cdots$$

This example illustrates the fact that even when the solutions are required to be positive there may be an infinite number of them.

Problems.

1. Divide 100 into two summands such that one is divisible by 7, the other by 11. (Euler.)

2. Required, such values of x and y in the indeterminate equation

$$7x + 19y = 1{,}921$$

that their sum $x + y$ may be the least possible. (From Barlow, *An Elementary Investigation of the Theory of Numbers*, etc., London, 1811.)

3. In the forest 37 heaps of wood apples were seen by the travelers. After 17 fruits were removed the remainder was divided evenly among 79 persons. What is the share obtained by each? (Mahaviracarya.)

4. Find two fractions having 5 and 7 for denominators whose sum is equal to $\frac{26}{35}$. (Barlow.)

5. A party of men and women have paid a total of 1,000 groschen. Every man has paid 19 groschen and every woman 13 groschen. What is the smallest number of persons the party could consist of? (Modified from Euler.)

6. How many different ways may £1,000 be paid in crowns and guineas? (Barlow.) [For non-English readers it may be recalled that one crown is 5 shillings, one pound 20 shillings, and one guinea 21 shillings.]

7. Solve problem 6 in Sec. 6–2.

8. Find a number that leaves the remainder 16 when divided by 39 and the remainder 27 when divided by 56.

6–4. Problems with several unknowns.

We turn now to those indeterminate problems in which there are more than two unknowns. There exists then a certain number of linear conditions; often the number of equations is just one less than the number of unknowns. The procedure is to eliminate some of the unknowns until one winds up with a single equation with two unknowns, which is the case we have just discussed. Most common is the case of two equations with three unknowns. In medieval times problems of this kind were known as *problema coeci*, a term of unknown origin.   The name probably refers to the fact that these problems often appeared in the form that a check should be paid by a certain number of people, as in the problem given by Bachet, for instance.   Sometimes they were also called *problema potatorum*, referring to drinkers and the mixing of wine, or also *problema virginum*, believed to have originated through certain problems given in terms of Greek mythology from the so-called *Palatine Anthology*.   The same problems are also reproduced in Bachet's collection.   In Euler's Algebra the *regula coeci* is illustrated first by the following example:

Thirty persons, men (x), women (y), and children (z), spend 50 thaler at an inn.   Each man pays 3 thaler, each woman 2 thaler, and each child 1 thaler.   How many persons were there in each category?

The equations are
$$x + y + z = 30, \qquad 3x + 2y + z = 50 \qquad (6\text{-}21)$$
By subtracting the first from the second, one obtains an equation with two unknowns
$$2x + y = 20$$
The positive solutions are obviously
$$x = \quad 0, \quad 1, \quad 2, \quad 3, \quad 4, \quad 5, \quad 6, \quad 7, \quad 8, \quad 9, 10$$
$$y = 20, 18, 16, 14, 12, 10, \quad 8, \quad 6, \quad 4, \quad 2, \quad 0$$
and from the first of equation (6–21), one finds the corresponding values
$$z = 10, 11, \quad 12, 13, 14, 15, 16, 17, 18, 19, 20$$
Thus there are 11 solutions.

For a less trivial example let us take problem 8 of Sec. 6–3, from Bachet.   From the second of equation (6–8) it is clear that z must be a number divisible by 3 so that one can write
$$z = 3z_1$$
When this is substituted in (6–8), the two equations become
$$x + y + 3z_1 = 41, \qquad 4x + 3y + z_1 = 40$$

When the last equation is multiplied by 3 and the first subtracted from it, one finds the equation with two unknowns

$$11x + 8y = 79$$

To find its general solution we proceed as previously and write

$$y = \frac{79 - 11x}{8} = 10 - x - \frac{3x + 1}{8}$$

Therefore the quotient

$$t = \frac{3x + 1}{8}$$

must be integral so that

$$8t = 3x + 1$$

or

$$x = 3t - \frac{t + 1}{3}$$

is an integer and

$$t = 3u - 1$$

Substituting this in the expressions for x and y, one obtains the general solution

$$x = 8u - 3, \qquad y = 14 - 11u$$

For positive solutions one must have

$$8u - 3 > 0, \qquad 14 - 11u > 0$$

or

$$\tfrac{3}{8} < u < 1\tfrac{2}{11}$$

This leaves as the only possibility $u = 1$ and $x = 5, y = 3$. From either one of the equations (6–8), it follows that $z = 33$. Thus there is a single solution to Bachet's problem.

Next let us consider the GI problem stated in (6–7). The second of these shows that z must be divisible by 10, hence $z = 10z_1$, and the equations become

$$x + y + 10z_1 = 100, \qquad 5x + 2y + z_1 = 100 \qquad (6\text{–}22)$$

When the first equation is multiplied by 2 and subtracted from the second, one finds

$$3x - 19z_1 = -100 \qquad (6\text{--}23)$$

This equation may be solved as before.   One writes

$$x = -\frac{19z_1 - 100}{3} = 6z_1 - 33 + \frac{z_1 - 1}{3}$$

Therefore

$$t = \frac{z_1 - 1}{3}$$

is an integer, and one obtains $z_1 = 3t + 1$ and

$$z = 30t + 10, \qquad x = 19t - 27$$

as the general solution of (6–23).   When it is substituted in the first equation (6–22), it follows that

$$y = 117 - 49t$$

To make all three numbers positive, one must have

$$19t - 27 > 0, \qquad 30t + 10 > 0, \qquad 117 - 49t > 0$$

or

$$t > 1\tfrac{8}{19}, \qquad t > -\tfrac{1}{3}, \qquad t < 2\tfrac{19}{49}$$

This is only possible for $t = 2$ so that one has the unique solution $x = 11$, $y = 19$, and $z = 70$, as already indicated.

Let us discuss another type of problem with three unknowns, which occurs later in the theory of congruences:

Find a number N that leaves the remainder 3 when divided by 11, the remainder 5 when divided by 19, and the remainder 10 when divided by 29.

The conditions are in this case

$$N = 11x + 3 = 19y + 5 = 29z + 10 \qquad (6\text{--}24)$$

Combining the two last conditions one has

$$19y = 29z + 5$$

One finds

$$y = z + \frac{10z + 5}{19}$$

and this shows that

$$\frac{10z + 5}{19} = \frac{5(2z + 1)}{19}$$

is integral. Since 5 is relatively prime to 19, it follows that

$$t = \frac{2z + 1}{19}$$

is integral. This gives in turn

$$z = 9t + \frac{t - 1}{2}$$

so that we write

$$u = \frac{t - 1}{2}$$

and find $t = 2u + 1$ and

$$z = 19u + 9, \qquad y = 29u + 14 \qquad (6\text{--}25)$$

These values for y and z give numbers N leaving the remainder 5 when divided by 19 and the remainder 10 when divided by 29. But one should also have the remainder 3 when the number is divided by 11. When the two first conditions in (6–24) are combined, one finds

$$11x = 19y + 2$$

Since the form of y is given by (6–25), it follows that x and u must be integers satisfying

$$11x = 551u + 268$$

This gives

$$x = 50u + 24 + \frac{u + 4}{11}$$

so that

$$v = \frac{u + 4}{11}$$

is integral.   We have, therefore, $u = 11v - 4$ and

$$x = 551v - 176$$

This in turn gives

$$N = 6{,}061v - 1{,}933$$

as the general form of the numbers with the desired residue properties.   The positive values of v give positive N, and the smallest
solution is obtained for $v = 1$ when $N = 4{,}128$.

Let us finally consider some problems in which the number of
unknowns is at least two greater than the number of equations.
In this case also, one can eliminate some of the unknowns and end
up with a single equation with several unknowns.   For instance
there may be two equations and four unknowns and one of them
may be eliminated to obtain a single equation with three unknowns.

Problem 5 in Sec. 6–2, which we quoted from the *Bija-Ganita*
of Bhaskara, is formally of this type.   But (6–5) may be written
simply as

$$y = z - 2x + 30$$

and the solution is trivial.   One can choose any integral positive x
and z arbitrarily greater than $2x - 30$ and find the corresponding
y.   In the solution given by Bhaskara the proportions $x : y : z$ of
the various prices are prescribed, and one obtains an ordinary
equation in a single variable.

A less trivial example from the same source, stated by Bhaskara
to be a problem from ancient authors, runs as follows:

Five doves are to be had for 3 drammas, seven cranes for 5, nine
geese for 7, and three peacocks for 9.   Bring 100 of these birds
for 100 drammas, for the prince's gratification.

When x, y, z, and t denote the numbers of doves, cranes, geese,
and peacocks, the equations are

$$x + y + z + t = 100, \qquad \tfrac{3}{5}x + \tfrac{5}{7}y + \tfrac{7}{9}z + 3t = 100 \quad (6\text{--}26)$$

When the first equation is multiplied by 3 and the second subtracted from it, one finds after the fractions have been cleared

$$189x + 180y + 175z = 15{,}750$$

Here one may apply the same reduction method as for equations with two unknowns. Solving for z, which has the smallest coefficient, it follows that

$$z = 90 - x - y - \frac{14x + 5y}{175}$$

so that

$$u = \frac{14x + 5y}{175}$$

is an integer. Thus we have the reduced equation

$$175u = 14x + 5y$$

or

$$y = 35u - 3x + \frac{x}{5}$$

This shows that

$$v = \frac{x}{5}$$

is integral, and proceeding backward one finds

$$x = 5v, \qquad\qquad y = 35u - 14v$$
$$z = 90 + 9v - 36u, \qquad t = u + 10$$

This represents the general solution of (6–26) in integers. The variables or parameters u and v are arbitrary integers. It should be noticed that in the case of one equation with two unknowns we obtained a general solution with one parameter, while in the case of three unknowns there will be two of them, as above.

Our problem requires that the solutions shall consist of positive numbers. This in general leads to a set of inequalities, which at times may be bothersome to analyze to find all possibilities. In our special case we must have $v > 0$ since $x > 0$, and consequently also $u > 0$ since $y > 0$. These conditions then insure that $x > 0$ and $t > 0$. From $y > 0$ and $z > 0$ one concludes

$$v > 4u - 10, \qquad v < \tfrac{5}{2}u$$

and therefore

$$\tfrac{5}{2}u > 4u - 10$$

or

$$u < \tfrac{20}{3}$$

Thus, there are the possibilities 1, 2, 3, 4, 5, 6, for u. Let us choose one of them, for instance, $u = 2$. When substituted in the general solution, this gives

$$x = 5v, \qquad y = 70 - 14v, \qquad z = 18 + 9v, \qquad t = 12$$

Here $y > 0$ so that v is limited, namely $v < 5$. The four possibilities $v = 1, 2, 3, 4$ correspond to the solution sets

	x	y	z	t
$v = 1$	5	56	27	12
$v = 2$	10	42	36	12
$v = 3$	15	28	45	12
$v = 4$	20	14	54	12

The first three of these are those actually given by Bhaskara. All the other possibilities for u and v may be investigated similarly. We shall leave it to the reader to derive *all* sets of solutions; it may only be stated that there are altogether 16 of them.

From Euler's *Algebra* we take our final example:

Someone buys 100 head of cattle for 100 thaler at the following prices: a steer, 10 thaler; a cow, 5 thaler; a calf, 2 thaler; and a sheep, $\tfrac{1}{2}$ thaler. How many did he buy of each kind?

Here are the equations.

$$x + y + z + t = 100, \qquad 10x + 5y + 2z + \tfrac{1}{2}t = 100$$

We multiply the last equation by 2 and subtract the first from it, giving

$$19x + 9y + 3z = 100$$

This in turn leads us to

$$z = 33 - 6x - 3y - \frac{x - 1}{3}$$

Hence, the only requirement is that

$$u = \frac{x - 1}{3}$$

shall be integral. One finds, therefore, the general integral solution in the form

$$x = 3u + 1, \qquad z = 27 - 19u - 3y, \qquad t = 72 + 16u + 2y$$

while y is arbitrary. Next one must analyze the conditions for positive solutions. Since

$$x = 3u + 1 > 0$$

one must have $u \geqq 0$ and naturally also $y > 0$. This is already sufficient to make $t > 0$. The remaining condition $z > 0$ leads to

$$19u + 3y < 27$$

Clearly u can take only the values $u = 0$ and $u = 1$. For $u = 0$, y can take the eight values $y = 1, 2, \cdots, 8$. For $u = 1$ there are only the possibilities $y = 1, y = 2$, and the two sets of solutions are

$$
\begin{array}{cccc}
x = 4, & y = 1, & z = 5, & t = 90 \\
4, & 2, & 2, & 92
\end{array}
$$

As these examples indicate, the determination of all positive solutions may often prove quite cumbersome. How complicated the matter may have appeared in earlier periods is evident from the following lament by an Arabic writer on the subject about A.D. 900. The title of his work is *The Book of Precious Things in the Art of Reckoning*, and the preface opens in this manner:

In the name of God, the compassionate and merciful. The writer is Shodja C. Aslam known by the name of Abu Kamil. I am familiar with a special kind of problems which circulate among high and low, among learned and among simple people, which they enjoy and which they find new and beautiful. But when one asks about the solution, one receives inaccurate and conjectural replies and they see in them neither principle nor rule. Many men, some distinguished and some humble, have asked me about problems in arithmetic and I replied to them for each separate problem with the single answer when there were no others. But often a problem had two, three, four and more answers, and often there was no solution. Indeed it happened to me in one problem which I solved that I found very many solutions. I considered the matter more penetratingly and came upon 2676 correct solutions. At this my surprize was great and I had the experience that when I told of the discovery, I was met with astonishment or was considered incompetent or those who did not know me had a false suspicion of me. Then I decided to write a book on the subject of such computations to facilitate the study and bring understanding nearer. This I have now begun and I shall declare the solutions for those problems which have several solutions and for those which have only one and for those which have none, all by means of an infallible method. Finally I shall treat a problem, which as I stated, has 2676 solutions. The suspicions will again disappear and my statement will be confirmed and the truth will show itself. It would carry too far if I should add more about the opinions which have been expressed to me in regard to the great number of solutions of this and similar problems.

To satisfy the curiosity of the reader it may be stated that the problem is of the same type as those considered in the last two examples. It has five unknowns and it leads to the equations

$$x + y + z + u + v = 100$$

$$2x + \tfrac{1}{2}y + \tfrac{1}{3}z + \tfrac{1}{4}u + v = 100$$

The reader is welcome to verify Abu Kamil's result.

To conclude, let us mention only a famous indeterminate problem known as the "cattle problem of Archimedes." Although the problem is ancient, it appears doubtful whether it ever had any connection with Archimedes. It is in poetic form and in it one is requested to find the number of head of cattle of various colors in the herds of the sun-god Helios as they graze the slopes of Sicily. The problem leads to seven equations in eight unknowns. These equations are quite simple and present no theoretical difficulty, but the numbers appearing in the solutions are enormous.

A later addition requires the solution of second-degree indeterminate equations.

Problems.

1. Solve problems 1, 2, and 3 in Sec. 6–2.

2. The following problems are quoted from the letters of the German mathematician Regiomontanus (1436–1476). In all of them, find the positive integral solutions.

(a) $97x + 56y + 3z = 16,047$

(b) $17x + 15 = 13y + 11 = 10z + 3$

(c) $23x + 12 = 17y + 7 = 10z + 3$

3. Write $\frac{17}{60}$ as a sum of three fractions with relatively prime denominators

4. Find the number of solutions in positive integers of the equation

$$5x + 11y + 13z = 2,000$$

and find the solution for which the sum $x + y + z$ is as small as possible.

5. In how many ways can one give change for (a) 25 cents, (b) 50 cents, (c) $1.

Bibliography

BACHET, C. G.: *Problèmes plaisans et delectables qui se font par les nombres.* Lyons, 1612. Many later editions.

CHACE, A. B., L. S. BULL, H. P. MANNING, and R. C. ARCHIBALD: *The Rhind Mathematical Papyrus*, 2 vols., Oberlin, Ohio, 1927, 1929.

COLEBROOKE, H. T.: *Algebra with Arithmetic and Mensuration from the Sanscrit of Brahmegupta and Bhaskara*, London, 1817.

EULER, L.: *Vollständige Anleitung zur Algebra*, St. Petersburg, 1770. Many English and French translations.

RANGACARYA, M.: *The Ganita-Sara-Sangraha of Mahaviracarya*, with English translation and notes, Madras, 1912.

CHAPTER 7

THEORY OF LINEAR INDETERMINATE PROBLEMS

7-1. Theory of linear indeterminate equations with two unknowns. After the many examples of linear indeterminate problems we have given in the preceding chapter, it is time to consider some of the more systematic aspects of their theory. The following result is essential in many applications of number theory:

THEOREM 7-1. When a and b are relatively prime, it is possible to find such other integers x and y that

$$ax + by = 1 \tag{7-1}$$

This may be stated slightly differently by saying that unity is a linear combination of a and b. The proof is an immediate application of Euclid's algorism. We suppose that $a > b$. To make the notation more systematic, we write $a = r_1$ and $b = r_2$ in stating the algorism:

$$\left.\begin{aligned}
r_1 &= q_1 r_2 + r_3 \\
r_2 &= q_2 r_3 + r_4 \\
\quad \cdot \quad \cdot \quad \cdot \quad \cdot \quad \cdot \quad \cdot \quad \cdot \quad \cdot & \\
r_{n-3} &= q_{n-3} \cdot r_{n-2} + r_{n-1} \\
r_{n-2} &= q_{n-2} \cdot r_{n-1} + 1
\end{aligned}\right\} \tag{7-2}$$

The last remainder is 1 since a and b are relatively prime. We shall now obtain a representation (7-1) by a stepwise process derived from (7–2). We begin at the bottom and write 1 as a linear combination of r_{n-2} and r_{n-1}

$$1 = r_{n-2} - q_{n-2}r_{n-1}$$

Here we substitute from the next to the last division

$$r_{n-1} = r_{n-3} - q_{n-3}r_{n-2}$$

and one obtains after rearrangement

$$1 = -q_{n-2}r_{n-3} + (1 + q_{n-2}q_{n-3})r_{n-2}$$

so that we have represented 1 as a linear combination of r_{n-3} and r_{n-2}. From the third last relation one introduces

$$r_{n-2} = r_{n-4} - q_{n-4}r_{n-3}$$

and in a similar manner one expresses 1 linearly by means of r_{n-4} and r_{n-3}. This process is continued until one arrives at a linear combination of $r_1 = a$ and $r_2 = b$ equal to 1, as the theorem requires.

Let us illustrate the procedure on the example $a = 109$ and $b = 89$. The algorism is

$$\underline{109} = \underline{89} \cdot 1 + \underline{20}$$
$$\underline{89} = \underline{20} \cdot 4 + \underline{9}$$
$$\underline{20} = \underline{9} \cdot 2 + \underline{2}$$
$$\underline{9} = \underline{2} \cdot 4 + \underline{1}$$

where the various remainders have been underscored to keep them separate from other figures occurring in the reductions. We begin by writing

$$1 = \underline{9} - 4 \cdot \underline{2}$$

and substitute

$$\underline{2} = \underline{20} - 2 \cdot \underline{9}$$

from the third division. This gives

$$1 = 9 \cdot \underline{9} - 4 \cdot \underline{20}$$

Here we substitute

$$\underline{9} = \underline{89} - 4 \cdot \underline{20}$$

from the second relation and obtain

$$1 = 9 \cdot \underline{89} - 40 \cdot \underline{20}$$

In the last step we use the first division and write

$$\underline{20} = \underline{109} - \underline{89}$$

so that we arrive at the desired representation

$$1 = 49 \cdot \underline{89} - 40 \cdot \underline{109}$$

It should be observed that when the algorism (7–2) was used to derive a solution of the equation (7–1), it was immaterial what the values of the remainders r_i were, as long as one had a set of relations of the type (7–2). This remark may be used to shorten the algorism in some cases, for instance, by taking least absolute remainders instead of least positive remainders. To illustrate we shall take $a = 249$ and $b = 181$, for which one finds

$$\underline{249} = \underline{181} \cdot 1 + \underline{68}$$

$$\underline{181} = \underline{68} \cdot 3 - \underline{23}$$

$$\underline{68} = \underline{23} \cdot 3 - 1$$

Again, in

$$1 = 3 \cdot \underline{23} - \underline{68}$$

we substitute the expression for $\underline{23}$ from the second relation and derive

$$1 = 8 \cdot \underline{68} - 3 \cdot \underline{181}$$

When $\underline{68}$ is eliminated in the same manner by means of the first relation, we obtain as a solution to our linear equation

$$1 = 8 \cdot \underline{249} - 11 \cdot \underline{181}$$

The work involved in the computation of a solution of the linear equation (7–1) may be reduced considerably by a systematic arrangement. Before we give the necessary proofs we shall illustrate the method on the example

$$1{,}027x + 712y = 1$$

Here the algorism is

$$\underline{1{,}027} = \underline{712} \cdot 1 + \underline{315}, \qquad 1 = -165 \cdot \underline{1{,}027} + 238 \cdot \underline{712}$$

$$\underline{712} = \underline{315} \cdot 2 + \underline{82}, \qquad 1 = 73 \cdot \underline{712} - 165 \cdot \underline{315}$$

$$\underline{315} = \underline{82} \cdot 3 + \underline{69}, \qquad 1 = -19 \cdot \underline{315} + 73 \cdot \underline{82}$$

$$\underline{82} = \underline{69} \cdot 1 + \underline{13}, \qquad 1 = 16 \cdot \underline{82} - 19 \cdot \underline{69}$$

$$\underline{69} = \underline{13} \cdot 5 + \underline{4}, \qquad 1 = -3 \cdot \underline{69} + 16 \cdot \underline{13}$$

$$\underline{13} = \underline{4} \cdot 3 + \underline{1}, \qquad 1 = 1 \cdot \underline{13} - 3 \cdot \underline{4}$$

$$1 = 0 \cdot \underline{4} + 1 \cdot \underline{1}$$

In the second column we have performed the substitutions required by our method, beginning at the bottom and proceeding successively to the solution on top. The lowest equation has been added as a supplement for a reason that will be clear instantly. To derive the rules of computation we shall establish, let us rewrite separately the coefficients in the right-hand column of equations above.

-165	238
73	-165
-19	73
16	-19
-3	16
1	-3
0	1

The two columns are the same except for one element at each end, so that we have the rule that the last coefficient in one equation becomes the first in the next. Furthermore the signs alternate. Consequently, to obtain our solution $x = -165$ and $y = 238$, it would suffice to compute the positive values of the numbers in the last column, and then give them the signs plus and minus alternatingly. This may be executed according to the following scheme:

$$\underline{1} \quad 238 = 165 \cdot \underline{1} + 73$$

$$\underline{2} \quad 165 = 73 \cdot \underline{2} + 19$$

$$\underline{3} \quad 73 = 19 \cdot \underline{3} + 16$$

$$\underline{1} \quad 19 = 16 \cdot \underline{1} + 3$$

$$\underline{5} \quad 16 = 3 \cdot \underline{5} + 1$$

$$\underline{3} \quad 3 = 1 \cdot \underline{3} + 0$$

In the first column we have written the quotients which occur in the divisions on the Euclid algorism. In the second column are the positive coefficients, each computed as indicated, by multiplying the corresponding quotient by the preceding coefficient and adding to the product the next preceding coefficient.

The proof of these rules is quite simple. Suppose that for some i we have found the relation

$$1 = -A_{i+1}r_i + A_i r_{i+1}$$

To eliminate the remainder r_{i+1} one must substitute

$$r_{i+1} = r_{i-1} - r_i q_{i-1}$$

from the algorism and one obtains

$$1 = A_i r_{i-1} - (A_i q_{i-1} + A_{i+1})r_i$$

This shows that the coefficient of r_{i-1} is the same as that of r_{i+1} in the preceding and also that one must put

$$A_{i-1} = -(A_i q_{i-1} + A_{i+1})$$

However, these are exactly the rules that were verified in the example above.

To give a final illustration of the scheme let us take the equation

$$1{,}726x + 1{,}229y = 1$$

The algorism is

$$
\begin{array}{rll}
1{,}726 = & 1{,}229 \cdot 1 + 497 & 639 \\
1{,}229 = & 497 \cdot 2 + 235 & 455 \\
497 = & 235 \cdot 2 + 27 & 184 \\
235 = & 27 \cdot 8 + 19 & 87 \\
27 = & 19 \cdot 1 + 8 & 10 \\
19 = & 8 \cdot 2 + 3 & 7 \\
8 = & 3 \cdot 2 + 2 & 3 \\
3 = & 2 \cdot 1 + 1 & 1 \\
& & 1
\end{array}
$$

The last column contains the computation of the successive coefficients by means of the quotients in the algorism. The two top numbers give the solution $x = -455$, $y = 639$, as is readily verified.

In this example, as well as in the preceding, we have assumed positive coefficients in the linear equation to be solved, and this is usually convenient. In order to find the solution of equations with negative coefficients, one need only observe that if x_0 and y_0 is a solution of (7–1), then

$$-x_0, y_0, \qquad x_0, -y_0, \qquad -x_0, -y_0$$

respectively are solutions of the equations

$$-ax + by = 1, \qquad ax - by = 1, \qquad -ax - by = 1$$

We shall turn next to the general linear equation in two unknowns

$$ax + by = c \qquad\qquad (7\text{–}3)$$

On the basis of the preceding analysis it is not difficult to find when such an equation can have integral solutions. Clearly, if the coefficients a and b in (7–3) have the greatest common divisor d, this factor must also divide c when there is to be an integral solution. Conversely, let d divide c. Then one can divide (7–3) by d and obtain an equation in which a and b are relatively prime. Let us suppose that this has been done.   It follows from theorem 7–1 that one can find integers x_0 and y_0 such that

$$ax_0 + by_0 = 1$$

When this equation is multiplied by c, one finds

$$a(cx_0) + b(cy_0) = c$$

and therefore

$$x = cx_0, \qquad y = cy_0$$

is a solution of (7–3).   To summarize we state:

THEOREM 7–2.   The necessary and sufficient condition for the equation

$$ax + by = c$$

to have a solution in integers is that the greatest common divisor of a and b divide c.

Examples.

1. The equation

$$114x + 312y = 28$$

has no solution in integers since the g.c.d. of 114 and 312 is 6 and this number does not divide 28.

2. The equation

$$208x + 136y = 120 \tag{7–4}$$

has solutions, since $(208, 136) = 8$ and this common divisor divides 120. After canceling by 8, we have the equation

$$26x + 17y = 15 \tag{7–5}$$

As indicated above, we first find a solution of the equation

$$26x + 17y = 1 \tag{7–6}$$

The algorism is

$$26 = 17 \cdot 1 + 9 \qquad 3$$
$$17 = 9 \cdot 1 + 8 \qquad 2$$
$$9 = 8 \cdot 1 + 1 \qquad 1$$
$$1$$

The last column contains the computation of the successive coefficients so that a solution of (7–6) is $x = 2$, $y = -3$. When these values are multiplied by 15, one obtains a solution $x = 30$, $y = -45$ of any of the two equivalent equations, (7–4) or (7–5).

3. To find a solution of the equation

$$1{,}726x + 1{,}229y = 3 \qquad (7\text{–}7)$$

we recall that we have already shown previously that the equation

$$1{,}726x + 1{,}229y = 1$$

has the solution

$$x = -445, \qquad y = 639$$

When these numbers are multiplied by 3, one arrives at the solution

$$x = -1{,}335, \qquad y = 1{,}917$$

for (7–7).

So far we have derived only one solution of the indeterminate equations we have studied. However, on the basis of one solution it is not difficult to find the general solution. We have already established that when there exists a solution of (7–3), the greatest common factor of a and b also divides c so that it may be canceled. We shall suppose in the following, therefore, that this has been done and a and b are relatively prime. When x_0 and y_0 form some particular solution of (7–3), one has

$$ax_0 + by_0 = c \qquad (7\text{–}8)$$

To find the general solution x and y, we subtract (7–8) from (7–3) and obtain

$$a(x - x_0) + b(y - y_0) = 0$$

which we prefer to write

$$a(x - x_0) = -b(y - y_0) \qquad (7\text{–}9)$$

This equation shows that the product of a and $x - x_0$ is divisible by b. Since a is relatively prime to b, we conclude that $x - x_0$ is divisible by b so that one can write

$$x - x_0 = tb$$

where t is some integer. When this is substituted back in (7–9) and b is canceled, one obtains

$$y - y_0 = -ta$$

Thus we have shown that one must have

$$x = x_0 + tb, \qquad y = y_0 - ta \qquad (7\text{--}10)$$

It may be verified directly that these values for x and y satisfy (7–3) regardless of the value of the integer t; hence in (7–10) we have the general form for the solution of the indeterminate equation.

By means of the general solution one can answer various questions about the existence of solutions with particular properties. Let us again consider some illustrations.

Examples.

1. Find the smallest positive integer that leaves the remainder 1 when divided by 1,000 and the remainder 8 when divided by 761.

The number must have the form

$$N = 1{,}000x + 1 = 761y + 8 \qquad (7\text{--}11)$$

so that

$$1{,}000x - 761y = 7 \qquad (7\text{--}12)$$

A solution of the equation

$$1{,}000x - 761y = 1$$

is found by the previous method to be

$$x = 121, \qquad y = 159$$

When multiplied by 7, it gives the solution to (7–12)

$$x = 847, \qquad y = 1{,}113$$

and the general solution becomes

$$x = 847 - 761t, \qquad y = 1{,}113 - 1{,}000t \qquad (7\text{--}13)$$

The smallest positive values for x and y are obtained when $t = 1$ and they are

$$x = 86, \qquad y = 113 \qquad (7\text{--}14)$$

The corresponding number asked for in the problem is, according to (7–11), $N = 86{,}001$.

In forming the general solution one could also have used the particular solution (7–14) so that

$$x = 86 - 761t, \qquad y = 113 - 1{,}000t$$

would be the general form, instead of (7–13). However, when t runs through all integers, the totality of solutions is, of course, the same in both forms. Let us also mention that since $-t$ runs through all integers when t does, one could also present the general solution as

$$x = 86 + 761t, \qquad y = 113 + 1{,}000t$$

2. Find the number of positive solutions to the equation

$$10x + 28y = 1{,}240$$

We first cancel the common factor 2 so that the equation reduces to

$$5x + 14y = 620$$

By inspection one sees that a particular solution of

$$5x + 14y = 1$$

is $x = 3$, $y = -1$. When this is multiplied by 620, it gives the particular solution $x = 1{,}860$, $y = -620$ for the previous equation. The general solution is therefore

$$x = 1{,}860 - 14t, \qquad y = -620 + 5t$$

To obtain positive solutions one must have

$$t < \frac{1{,}860}{14} = 132\frac{6}{7}, \qquad t > 124$$

or

$$132 \geqq t \geqq 125$$

so that there are 8 positive solutions. For $t = 124$ one finds $y = 0$.

One can also look at this theory of linear indeterminate equations from a geometric point of view. In analytic geometry, an equation

$$ax + by = c$$

represents a straight line. The points (x, y) in the plane whose coordinates x and y are integers are called *lattice points*. To solve the linear equation in integers means, therefore, to determine those lattice points that lie on the line. The general form of the integral solutions, as we have found it, shows that if (x_0, y_0) is a solution, then there are lattice points on the line for all the abscissas

$$x_0, \qquad x_0 \pm a, \qquad x_0 \pm 2a, \qquad \cdots$$

This means that the lattice points that represent solutions lie at

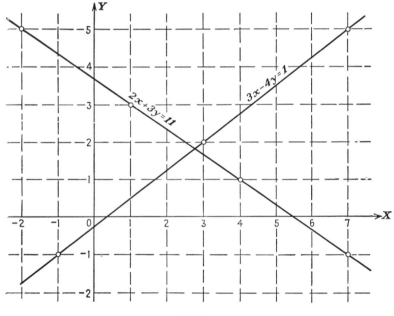

Fig. 7–1.

even intervals on the line with abscissas differing by a, and similarly, with ordinates y differing by b.

The situation has been illustrated in the figures representing the two lines

$$2x + 3y = 11, \qquad 3x - 4y = 1$$

These two illustrations clarify another fact. One can write the equation for the straight line in the form

$$y = -\frac{a}{b}x + \frac{c}{b}$$

Therefore the slope of the line is positive when a and b have different signs, as in the second example. But when the slope is positive, the line must have an infinitely long portion in the first quadrant, in which both x and y are positive, so that when a and b have different signs, there will be an infinite number of positive integral solutions, provided of course that there are any at all. On the other hand, when a and b have the same sign, for instance, both positive as in the first example, the line, if it goes through the first quadrant, can have only a finite portion in this part of the plane. In this case there can at most be a finite number of positive solutions.

Problems.

1. Find the general integral solution and also the positive solutions for each of the following equations:

(a) $39x - 56y = 11$

(b) $311x + 712y = 1{,}300$

(c) $7x + 13y = 71$

(d) $39x - 119y = 49$

(e) $170x - 445y = 625$

2. Find the number of positive solutions of the equations

(a) $33x + 41y = 1{,}946$

(b) $31x - 7y = 2$

(c) $3x + 11y = 1{,}000$

7–2. Linear indeterminate equations in several unknowns. So far we have discussed the theory of linear indeterminate equations in two unknowns in considerable detail. We turn next to such equations with several unknowns. These equations, as we saw in the examples, will appear in problems in which the number of unknowns is one or more greater than the number of equations.

The basic result for several unknowns is quite analogous to the main result for two unknowns as we expressed it in theorem 7–1 In a slightly different form this theorem may be stated: When d is the greatest common divisor of two numbers a and b, one can find two other integers x and y such that

$$ax + by = d \qquad (7\text{--}15)$$

For several unknowns we shall derive the corresponding theorem:

THEOREM 7–3. Let

$$a_1, a_2, \ldots, a_n$$

be a set of integers with the greatest common divisor

$$d = (a_1, a_2, \cdots, a_n)$$

Then one can find such integers

$$x_1, x_2, \ldots, x_n$$

that

$$a_1x_1 + a_2x_2 + \cdots + a_nx_n = d \qquad (7\text{--}16)$$

To prove this theorem one can proceed in various ways. We shall prefer to use the induction method, and to this end we need to observe that the result is obviously true when there is only one number a_1 (then $d = a_1$ and $x_1 = 1$). Also we have just mentioned that the theorem is true when there are two numbers a_1 and a_2. The induction consists in supposing that the theorem is true when there are $n - 1$ numbers a_i and applying this to prove it for n numbers. We shall denote by

$$d_{n-1} = (a_1, \cdots, a_{n-1})$$

the g.c.d. of the $n - 1$ first numbers. According to our assumption, we can find $n - 1$ numbers

$$y_1, \ldots, y_{n-1}$$

such that

$$a_1y_1 + \cdots + a_{n-1}y_{n-1} = d_{n-1} \qquad (7\text{--}17)$$

But let us recall from the properties of the g.c.d. that d, the g.c.d. of all a_i's is also the g.c.d. of d_{n-1} and a_n

$$d = (d_{n-1}, a_n)$$

This means that one can find two integers t and x_n so that

$$d_{n-1}t + a_n x_n = d$$

When we put in the value of d_{n-1} from (7–17), we arrive at the relation

$$a_1 y_1 t + \cdots + a_{n-1} y_{n-1} t + a_n x_n = d$$

and this is exactly of the form (7–16) when we write

$$x_1 = y_1 t, \qquad x_2 = y_2 t, \cdots, x_{n-1} = y_{n-1} t$$

This proof has the advantage of providing a fairly simple method of computing a solution to (7–16).

Example.

The three numbers

$$a_1 = 100, \qquad a_2 = 72, \qquad a_3 = 90$$

have the g.c.d. $d = 2$, and we shall find numbers x_1, x_2, and x_3 such that

$$100x_1 + 72x_2 + 90x_3 = 2 \qquad (7\text{–}18)$$

The g.c.d. of 100 and 72 is $d_2 = 4$ so that we begin by solving the equation

$$100y_1 + 72y_2 = 4$$

or

$$25y_1 + 18y_2 = 1$$

By the usual method one finds a solution $y_1 = -5$, $y_2 = 7$. In the next step we solve

$$4t + 90x_3 = 2$$

or

$$2t + 45x_3 = 1$$

By inspection one sees that $x_3 = 1$, $t = -22$ satisfies this equation. By multiplying y_1 and y_2 by -22 one obtains as a solution of (7–18)

$$x_1 = 110, \qquad x_2 = -154, \qquad x_3 = 1 \qquad (7\text{–}19)$$

Analogously, as in the case of two unknowns, one can prove further:

THEOREM 7–4. The necessary and sufficient condition for the equation

$$a_1 x_1 + \cdots + a_n x_n = c \qquad (7\text{–}20)$$

to be solvable in integers $x_1, \ldots, x_n$ is that c be divisible by the greatest common divisor of all numbers a_i.

The proof is simple. Since d, the g.c.d. of all numbers a_i, divides each of them, it must also divide c if there is to be an integral solution. On the other hand, when c is divisible by d, one can divide all terms in (7–20) by d and obtain a new equation in which the coefficients of the unknown have 1 for their g.c.d. Let us suppose that this reduction has already been carried out in (7–20). Then according to theorem 7–3 one can find such integers $y_1, \ldots, y_n$ that

$$a_1 y_1 + \cdots + a_n y_n = 1$$

When this relation is multiplied by c, one has

$$a_1(cy_1) + \cdots + a_n(cy_n) = c$$

so that the multiples $x_i = cy_i$ give a solution of (7–20).

Examples.

1. Let us take first the equation

$$100x_1 + 72x_2 + 90x_3 = 11$$

Since the g.c.d. of 100, 72, and 90 is $d = 2$, and since this number does not divide $c = 11$, the equation has no integral solution.

2. On the other hand, the equation

$$100x_1 + 72x_2 + 90x_3 = 6 \tag{7–21}$$

does have solutions according to our criterion since $d = 2$ divides $c = 6$. To find one of them we divide (7–21) by 2 and obtain

$$50x_1 + 36x_2 + 45x_3 = 3 \tag{7–22}$$

and then solve the equation

$$50y_1 + 36y_2 + 45y_3 = 1$$

This, however, is the same as (7–18) divided by 2, so that (7–19) gives a solution to it. When these numbers are multiplied by 3, one finds

$$x_1 = 330, \qquad x_2 = -462, \qquad x_3 = 3$$

to be a solution of (7–21) and (7–22).

When there were only two unknowns, it was fairly simple to derive the general solution as soon as one knew a particular set of values satisfying the equation. For several unknowns, the situation is more complicated, as we have already seen in the examples in Chap. 6. We indicated there how one could find the general solution by a series of reductions, and this method is probably the best available. Let us illustrate it once more by deriving the general solution of (7–22). We begin by writing

$$x_2 = \frac{3 - 50x_1 - 45x_3}{36} = -x_1 - x_3 + \frac{3 - 14x_1 - 9x_3}{36}$$

Therefore

$$t = \frac{3 - 14x_1 - 9x_3}{36}$$

is an integer. When this expression is solved for x_3, it follows that

$$x_3 = -4t - 2x_1 + \frac{3 + 4x_1}{9}$$

Consequently

$$u = \frac{3 + 4x_1}{9}$$

is integral and

$$x_1 = 2u + \frac{u - 3}{4}$$

This gives finally that

$$v = \frac{u - 3}{4}$$

is integral and $u = 4v + 3$. When this is substituted, one finds as the general solution

$$x_1 = 9v + 6$$

$$x_2 = 5t + 5v + 3$$

$$x_3 = -4t - 14v - 9$$

where v and t are arbitrary integers.

Problems.

1. Find an integral solution to each of the equations:

    (a) $31x + 49y - 22z = 2$

    (b) $120x + 84y + 144z = 22$

2. Find the general solution and the number of positive solutions of each of the equations:

    (a) $31x + 49y + 22z = 1,000$

    (b) $102x + 311y + 202z = 10,000$

7–3. Classification of systems of numbers. In mathematics one deals with many systems of numbers, characterized by various properties. It has gradually become evident that certain systems are of particular importance, namely, those that reproduce themselves, or, as one prefers to say, are *closed*, under some or all of the four arithmetic operations, addition, subtraction, multiplication, and division. For such systems there has come into use fairly recently a nomenclature we shall now explain. Although we shall use these terms only incidentally in subsequent chapters, they are of such importance and common occurrence even in fairly elementary mathematical writings that the reader should be familiar with them.

Let S be a set of numbers of any kind. We shall say that S is *closed under addition* if, for any two numbers a and b in S, their sum $a + b$ is also a number in S. It follows immediately that the sum of three, four, or any finite number of elements in S will again belong to S. Let us illustrate this definition by some examples.

Example 1. The set of all natural numbers $1, 2, 3, \ldots$ is closed under addition.

Example 2. The even integers form a system closed under addition. The odd integers are not closed under addition.

Example 3. All positive real numbers form a system closed under addition.

Example 4. The sets of all integers, all rational numbers, all real numbers, and all complex numbers are closed under addition.

The closure with respect to subtraction is defined analogously. We say that a set of numbers S is *closed with respect to subtraction*, when the difference $a - b$ of any two of its numbers again belongs to S. Such a set is called a *modul*.

Example 1. The integers $0, \pm 1, \pm 2, \ldots$ form a modul, but the natural numbers $1, 2, 3, \ldots$ do not.

Example 2. The even integers form a modul.

Example 3. The sets of all rational numbers, all real numbers, and all complex numbers are moduls.

Example 4. All purely imaginary numbers ib form a modul.

We shall now show that a modul has the following properties:

THEOREM 7–5. (*a*) A modul contains 0.

(*b*) When a modul contains a it contains $-a$.

(*c*) A modul is closed with respect to addition.

Proof: When a is an element in a modul S, the difference $a - a = 0$ is in S. Consequently $0 - a = -a$ is in S, and therefore also

$$a + b = a - (-b)$$

when a and b are in S.

As a consequence a modul is sometimes defined as a system closed under both addition and subtraction.

Since number theory deals primarily with the integers, we are interested in finding all moduls consisting only of integers. Clearly one type of such moduls may be obtained simply by taking all multiples $k \cdot a$ of some integer a because the sum and difference of two multiples is again a multiple.

$$k_1 a \pm k_2 a = (k_1 \pm k_2)a$$

It is remarkable, however, that *all* moduls of integers are of this kind. This is a consequence of the following theorem, which we shall now prove:

THEOREM 7–6. Any modul M containing only integers consists of all multiples of the greatest common divisor of the numbers in M.

We remark first that if a is some integer in a system that is closed under addition, every multiple $k \cdot a$ is also in the system since it is the sum of a taken k times.   Second, let a and b be some integers contained in a modul M.   Then $k \cdot b$ is in M for any integral k and therefore also any difference

$$r = a - kb$$

In particular, we conclude that when a is divided by b, the remainder r is in the modul.

To prove theorem 7–6 we select the smallest positive integer d in M.   Such an integer must exist except in the trivial case where M consists of the single number 0.   All positive and negative multiples of d are in M, and they exhaust M.   If, for example, m is some integer in M, we divide m by d

$$m = kd + r, \qquad d > r \geqq 0$$

and r, as we remarked, also belongs to M.   But since d was the smallest positive integer in M, this is possible only when $r = 0$, and $m = kd$ is a multiple of d.   Obviously d is the g.c.d. of the numbers in M.

It is of interest to connect the properties of moduls with the linear indeterminate equations.   We shall use the result expressed in theorem 7–6 to derive the basic theorem 7–3 for equations. Here were given n numbers

$$a_1, a_2, \ldots, a_n$$

All numbers of the form

$$x = x_1 a_1 + \cdots + x_n a_n \qquad (7\text{--}23)$$

with integral x_i's will form a modul M, since the sum and difference of two such numbers will be of the same kind. M consists of integers and it is not difficult to find their g.c.d.   All numbers a_i belong to M because one can write, for instance,

$$a_1 = 1a_1 + 0a_2 + \cdots + 0a_n$$

But

$$d = (a_1, \cdots, a_n)$$

divides all a_i's and consequently all numbers (7–23) so that d is the g.c.d. of the numbers in M.   From theorem 7–6 we know that M consists of all multiples of d; in particular, d belongs to M so that it is also of the form (7–23)

$$d = x_1a_1 + \cdots + x_na_n$$

with suitable x_i. This is the content of theorem 7–3.

A set of numbers S is *closed under multiplication* when the product $a \cdot b$ of any two of its elements a and b is again in S. One concludes that then any finite number of elements in S has a product belonging to S. A set closed under multiplication is sometimes called a *ray*. Among the examples let us mention:

Example 1. The natural numbers as well as the integers.

Example 2. The even integers and also the odd integers.

Example 3. The rational numbers, the real numbers, and the complex numbers.

Example 4. The real numbers between 0 and 1.

A system of numbers that is closed under addition, subtraction, and multiplication is called a *ring*. From theorem 7–5 we see that the specific mention of addition in this definition is superfluous, nevertheless it is usually included in the statement. One can also say that a ring is a modul that is closed under multiplication. Among the many examples are:

Example 1. The integers form a ring.

Example 2. The even numbers form a ring.

Example 3. The rational, real, and complex numbers define rings.

Example 4. All complex numbers $a + ib$, where a and b are integral, form a ring.

Example 5. All numbers of the form $a + b\sqrt{2}$ where a and b are integers form a ring. To verify this, we observe that the difference of two such numbers is of the same form, and furthermore

$$(a + b\sqrt{2})(c + d\sqrt{2}) = ac + 2bd + (bc + ad)\sqrt{2}$$

Example 6. A similar argument shows that for a fixed integer D all numbers $a + b\sqrt{D}$ with integers a and b form a ring.

We have shown in theorem 7–6 that a modul consisting of integers must consist of all multiples $k \cdot a$ of some number a.   It is of interest to note that such a modul is also a ring, since the product of two such multiples

$$(k_1 a)(k_2 a) = (k_1 k_2 a)a$$

is another multiple of a.

We turn finally to the sets of numbers that are *closed under division*.   Such a set contains the quotient a/b of any two of its elements, provided $b \neq 0$.

Example 1.   The set of positive real numbers is closed under division.

Example 2.   The set consisting of the single number 0 may be considered to be closed under division.   This is somewhat improper of course, since by the definition of such systems the division by 0 was excluded.

As a result of the analogy we have already mentioned, between the laws for addition and those for multiplication, one can derive a theorem analogous to theorem 7–5.

THEOREM 7–7.   A set of numbers S, consisting not only of 0, that is closed under division, must have the three properties:

> (*a*) S contains 1.
> (*b*) When $a \neq 0$ is in S, so is a^{-1}.
> (*c*) S is closed with respect to multiplication.

Proof: When $a \neq 0$ is in S, the quotient $a/a = 1$, hence the quotient $a^{-1} = \dfrac{1}{a}$ must also belong to S.   Consequently, for any b in S,

$$ab = \frac{b}{a^{-1}}$$

is in S.

A system that does not include 0 and is closed under division is called a *multiplicative group*.   A simple example is the set of all positive and negative powers

$$1, a, a^2, \ldots, a^{-1}, a^{-2}, \ldots$$

of some number $a \neq 0$.

We now come to the last definition of this kind: A *field* is a set of numbers that is closed under all four arithmetic operations, addition, subtraction, multiplication, and division. From theorems 7–5 and 7–7, we see that in this definition the inclusion of addition and multiplication is superfluous. One could also have defined a field as a system that is a modul and in which the elements after exclusion of 0 form a multiplicative group. Again we illustrate by some examples:

Example 1. The most trivial case of a field would be the number 0 alone. This case is so exceptional that it is ordinarily excluded and not counted as a field.

Example 2. The rational, the real, and the complex numbers all form fields.

Example 3. All numbers of the form $a + b\sqrt{2}$ with rational a and b form a field. To verify this, one observes that the difference of two such numbers belongs to the set. Furthermore, the reduction

$$\frac{a + b\sqrt{2}}{c + d\sqrt{2}} = \frac{ac - 2bd}{c^2 - 2d^2} + \frac{bc - ad}{c^2 - 2d^2}\sqrt{2}$$

shows that the quotient can be written in the same form.

Example 4. Clearly the preceding example can be extended. Let D be some fixed, positive or negative integer that is not a square, so that $\sqrt{D}$ is not rational. As before, one can show that the numbers $a + b\sqrt{D}$ with rational a and b form a field. Such fields are called *quadratic fields*.

There are many other rings and fields, some of great importance in number theory and algebra. For our purposes the examples given above are quite sufficient. We shall conclude these remarks with a theorem that shows that the rational field in a sense is the smallest possible field:

THEOREM 7–8.   Every field contains the rational field.

Proof: From theorem 7–7 one concludes that any field contains the number 1.   Since the sum of any number of these 1's is in the field, all natural numbers are in the field, and so are all integers according to theorem 7–6.   Since every rational number is the quotient of integers, our theorem is proved.

Problems.

1. Under which arithmetic operations are the following sets of numbers closed:

    (*a*) The real numbers ≥ 1
    (*b*) The numbers of the form $1/n$ where n is integral
    (*c*) The numbers of the form $n/2$ where n is integral
    (*d*) The complex numbers $a + ai$ where a is integral

2. Show that all fractions whose denominators are powers of 2 form a ring.
3. Prove that all numbers of the form

$$a + b\,\frac{1 + \sqrt{5}}{2}$$

with integral a and b form a ring.

Bibliography

ALBERT, A. A.: *College Algebra*, McGraw-Hill Book Company, Inc., New York, 1946.

BIRKHOFF, G., and S. MACLANE: *A Survey of Modern Algebra*, The Macmillan Company, New York, 1941.

MACDUFFEE, C. C.: *An Introduction to Abstract Algebra*, John Wiley & Sons, Inc., New York, 1940.

CHAPTER 8

DIOPHANTINE PROBLEMS

8-1. The Pythagorean triangle. Among the many classical Greek schools of mathematics and philosophy the Pythagorean was the oldest and most venerable. Pythagoras was born around 570 B.C., according to the best estimates. Tradition has it that he came from the island of Samos and traveled widely before he established his school in Crotona in Southern Italy. In Egypt and Babylonia he absorbed the lore of mysticism and also learned the laws of numbers and geometry. Pythagoras must have had considerable personal charm and conviction; his school became a fashionable center and attracted large numbers of students—some as auditors while the more qualified were eligible to be initiated in an inner circle of advanced and mystical learning. The school continued actively for at least a century after the death of Pythagoras, and it preserved its esoteric character as a society of fellows searching for the divine laws of knowledge. The extent of Pythagoras's own creative contributions to the science of mathematics is difficult to estimate, both because his doctrines were propounded only in his lectures and transmitted orally without permanent records, and also because his disciples generally effaced their personal roles by ascribing their discoveries to the founder of the school. The strands of mathematical history are further snarled because later Greek writers almost traditionally ascribed the early mathematical discoveries to the Pythagoreans when their provenance was not otherwise known.

The *Pythagorean theorem* states that in a right triangle the square constructed on the hypotenuse is equal to the sum of the squares on the two legs. When c is the length of the hypotenuse and a and b the lengths of the two other sides, the theorem becomes

$$a^2 + b^2 = c^2 \qquad (8\text{--}1)$$

This result was certainly known to the Pythagoreans and they may have been the first to give a satisfactory proof.

One of the simplest cases of the theorem occurs when the sides are

$$a = 3, \qquad b = 4, \qquad c = 5$$

since

$$3^2 + 4^2 = 5^2 \qquad (8\text{--}2)$$

The knowledge of this particular case has been widespread. One finds it in the earliest Chinese and Hindu works, together with other examples where the sides may be represented by integers, for instance,

$$a = 5, \qquad b = 12, \qquad c = 13$$
$$a = 8, \qquad b = 15, \qquad c = 17$$

In view of the particular interest of the Pythagoreans in relations that could be expressed in whole numbers, it would appear natural that they should have investigated the problem of finding right triangles with integral sides. There exists, according to Proclus, a much later writer, a formula for a certain type of solution to the equation (8–1), which he ascribes to Pythagoras. The formula is

$$a = 2n + 1, \qquad b = 2n^2 + 2n, \qquad c = 2n^2 + 2n + 1 \quad (8\text{--}3)$$

where n is any integer. It may be verified by substitution that the values (8–3) actually satisfy the relation (8–1). For $n = 1, 2, 3$, one finds the following triplets of solutions:

$$(3, 4, 5), \qquad (5, 12, 13), \qquad (7, 24, 25)$$

and this may be continued to give infinitely many others. Pythagoras's solution, as one sees, has the special property that the hypotenuse exceeds the larger leg by one. Another special solution is ascribed to Plato. The first general solution of the Pythagorean problem is found in the tenth book of Euclid's *Elements*, shrouded in geometric terms according to the custom of Greek mathematics at the time.

Let us examine how one may arrive at the general solution of the equation (8–1) in positive integers. We remark that the restriction to integers is not essential; if any rational solution had been found one could write the three numbers on a common denominator

$$a = \frac{a_1}{m}, \qquad b = \frac{b_1}{m}, \qquad c = \frac{c_1}{m}$$

and it would follow that

$$a_1{}^2 + b_1{}^2 = c_1{}^2$$

would be an integral solution.

It is sufficient to find the *primitive* integral solutions of the equation, *i.e.*, those solutions in which there is no factor common to a, b, and c, because if such a factor did occur the equation could be canceled by d^2. But for a primitive solution any *pair* of two of the numbers a, b, and c must be relatively prime. If for instance a and b had a common factor e, the left-hand side in (8–1), and hence also c^2, would be divisible by e^2. But then c is divisible by e, contrary to the assumption that the solution was primitive.

The next step is to see that in a primitive solution a, b, and c, the numbers a and b cannot both be odd. This is a consequence of theorem 2–1. Since the square of an odd number leaves the remainder 1 when divided by 4, it would follow that if a and b were both odd, the left-hand side in (8–1), hence also c^2, would leave the remainder 2 when divided by 4, contrary to the theorem just mentioned. We suppose now that the notation is taken such that a is even; consequently b and c are odd since there are no common factors. The equation (8–1) may be written

$$a^2 = c^2 - b^2 = (c + b)(c - b)$$

According to the preceding statements both sides are divisible by 4, and when this factor is divided out, one has

$$\left(\frac{a}{2}\right)^2 = \frac{c + b}{2} \cdot \frac{c - b}{2} \tag{8–4}$$

Here the two integral factors on the right are relatively prime, because any common factor d would divide both the sum and the difference of them. But since

$$\frac{c + b}{2} + \frac{c - b}{2} = c$$

$$\frac{c + b}{2} - \frac{c - b}{2} = b$$

and b and c are relatively prime, one must have $d = 1$.

When the two numbers on the right in (8–4) are relatively prime, their prime factors are different, and their product cannot be a square unless each of them is a square. We can put, therefore,

$$\frac{c + b}{2} = u^2, \qquad \frac{c - b}{2} = v^2$$

and from this we obtain by substitution in (8–4)

$$a = 2uv, \qquad b = u^2 - v^2, \qquad c = u^2 + v^2 \qquad (8\text{–}5)$$

To ensure that this solution is actually primitive, we observe that any common factor of b and c must divide the sum and difference of these numbers. But since

$$c + b = 2u^2, \qquad c - b = 2v^2$$

and since u and v are relatively prime, the only possible common factor is 2. This factor is excluded when one of the numbers u and v is odd and the other even.

From the general primitive solution (8–5) where u and v are integers subject to the conditions just mentioned, one finds the general integral solution of the Pythagorean equation (8–1) by multiplying by an arbitrary integer. The general rational solution is obtained by multiplication of (8–5) by a rational number. A little later on, however, in connection with a problem by Diophantos, we shall need the general rational solution, and it is convenient to have the formulas in a slightly different form. Let us divide both sides of a, b, and c in (8–5) by v^2 so that

$$\frac{a}{v^2} = 2\frac{u}{v}; \qquad \frac{b}{v^2} = \left(\frac{u}{v}\right)^2 - 1, \qquad \frac{c}{v^2} = \left(\frac{u}{v}\right)^2 + 1$$

There exists, therefore, to the given solution (8–5), a proportional rational solution

$$a_1 = 2t, \qquad b_1 = t^2 - 1, \qquad c_1 = t^2 + 1$$

where we have put $t = u/v$. When these values are multiplied by some rational number, one obtains the general solution

$$a_0 = 2tr, \qquad b_0 = (t^2 - 1)r, \qquad c_0 = (t^2 + 1)r \qquad (8\text{–}6)$$

where r and t are arbitrary rationals.

Some of the primitive integral solutions in the smallest numbers may be obtained from (8–5).

		a	b	c
$u = 2,$	$v = 1$	4	3	5
$u = 3,$	$v = 2$	12	5	13
$u = 4,$	$v = 1$	8	15	17
$u = 4,$	$v = 3$	24	7	25

Extensive tables of integral Pythagorean triangles have been computed; one, for instance, by A. Martin[1] gives all primitive triangles for which the hypotenuse does not exceed 3,000.

There are a great number of questions one may ask in regard to the Pythagorean triangles, and through the centuries they have been the source of many number-theory problems. A simple one suggested by the special Pythagorean solution (8–3) is: When does the hypotenuse differ from one of the legs by 1? One cannot have

$$c - b = 1$$

[1] *Proceedings*, Fifth International Mathematical Congress, Cambridge, 1912.

because when the values (8–5) are substituted, one finds the impossible equation in integers

$$2v^2 = 1$$

The other possibility

$$c - a = 1$$

leads to

$$u^2 + v^2 - 2uv = (u - v)^2 = 1$$

so that $u = v + 1$. When this is substituted in (8–5), one obtains

$$a = 2v^2 + 2v, \qquad b = 2v + 1, \qquad c = 2v^2 + 2v + 1$$

This is, however, the Pythagorean solution (8–3) when the a and b are interchanged in the notation.

Other problems have been discussed, for instance, the determination of all integral triangles in which the legs differ by 1, of triangles with special properties of the perimeter or area, of the number of right triangles with a given side, and so on.

Problems.

1. Find all integral Pythagorean triangles in which one leg differs from the hypotenuse by 2 or 3.

2. Find all integral Pythagorean triangles with hypotenuse not exceeding 50.

3. Try to find the general solution in integers of the equations

(a) $2x^2 + y^2 = z^2$

(b) $3x^2 + y^2 = z^2$

by the method used to solve the Pythagorean triangle.

8–2. The Plimpton Library tablet. Our brief sketch of the early history of the Pythagorean problem would have covered the main facts until quite recently. However, in the last decade or two new light has been thrown on the whole beginning of mathematics through a deeper understanding of the extent of Babylonian mathematics. The existence of early mathematical results among the Babylonians had long been known or suspected, partly through statements in Greek sources, partly through scattered cuneiform texts. It had also been known that the larger Babylonian collections, particularly those at the British Museum and the Louvre abroad, and at Yale and the University of Pennsylvania in this country, possessed a considerable number of undeciphered cuneiform tablets of unusual types. They often contained columns of figures, and for that reason they had in some cases been summarily classified as "commercial accounts." Recent investigations, particularly those by Neugebauer and Thureau-Dangin, have revealed that they are actually mathematical tables and texts. With this fact as a key, the reading was not difficult.

Through this rich mine of source material we have gained a surprisingly intimate view of Babylonian mathematics and its role in society. The tablets cover a period from 2000 B.C. to 200 B.C., but even the oldest ones contain methods that are quite advanced so that the origin of such methods may safely be placed at a considerably earlier period. The cuneiform tablets give calculations of areas and volumes, to a large extent as practical problems arising in connection with surveying and construction, digging of dikes, and building of walls. Other problems contain questions regarding the computation of simple and compound interest or division of estates according to rather involved laws and customs. One also finds theoretical problems, some of them strikingly like those given in elementary mathematics today. It is evident that the Babylonians were familiar with problems that led to second-degree equations, and the square roots that occurred in their solution were determined much as they are today, namely, by means of tables. As a whole, Babylonian mathematics made systematic and extensive use of numerical tables. Numerous multiplication tables, tables of inverses, squares, and square roots, tables of powers of a number, tables for finding the circumference of a circle, and several other types have been preserved and may be found in the Babylonian collections.

Let us dwell for a moment on the tables of inverses, which are particularly common among the Babylonian tablets. The operation of division appears to have been a relatively difficult one to master in the development of arithmetic in all countries. In medieval Europe a man capable of performing long division was probably more rare than a man with a Ph.D. at present. The simple procedure of successive duplications and mediations, which we described in Chap. 2, has been widely used. The Babylonians used tables of inverses to reduce division to multiplication. To find the value of the fraction a/b, one wrote it as a multiplication $a \cdot 1/b$ where the value of $1/b$ could be found as a sexagesimal fraction in the tables.

In constructing these tables of sexagesimal inverses, one runs into the same trouble as in the ordinary expansion in decimal fractions, namely, that some expansions are infinite and do not break off, as for instance

$$\tfrac{1}{3} = 0.333 \cdots, \qquad \tfrac{1}{7} = 0.142857\ 142857 \cdots$$

In the most common tables this difficulty is circumvented by including only numbers whose inversions have a finite, in fact, a rather short, sexagesimal expansion. According to Neugebauer the standard type of table of reciprocals usually contains the following pairs:

a	a^{-1}	a	a^{-1}	a	a^{-1}
2	30	13	3, 45	45	1, 20
3	20	18	3, 20	48	1, 15
4	15	20	3	50	1, 12
5	12	24	2, 30	54	1, 6, 40
6	10	25	2, 24	1	1
8	7, 30	27	2, 13, 20	1, 4	56, 15
9	6, 40	30	2	1, 12	50
10	6	32	1, 52, 30	1, 15	48
12	5	36	1, 40	1, 20	45
15	4	40	1, 30	1, 21	44, 26, 40

We have preserved the sexagesimal notation in the table. In checking the figures, the reader should recall that the Babylonians used no decimal sign to indicate where the units begin, so that, for instance, 60 may denote not only this figure but also 1 or 60^2. One observes that in the table, entries like 7, 11, 13, 14, and so on, which would give infinite expansions, have been omitted.

Although they are comparatively rare, there also exist tables that within their limits give the reciprocals of *all* numbers without exception. For numbers with an infinite sexagesimal expansion,

Fig. 8–1. Table of inverses. (*Courtesy of Yale Babylonian Collection.*)

one obtains a satisfactory approximation by breaking it off after a certain number of places as in our ordinary numerical tables. The Yale Babylonian Collection, which is particularly rich in mathematical source material, contains one tablet (YBC 10,529) which in its preserved part gives the reciprocals of all numbers between 58 and 80 with great accuracy. (Fig. 8–1.)

It is not difficult to determine which numbers have a finite sexagesimal expansion

$$\frac{a}{b} = a_0 + \frac{a_1}{60} + \cdots + \frac{a_r}{60^r} = (a_0, a_1, \cdots, a_r)$$

Clearly this can occur only when the fraction can be written in the form

$$\frac{a}{b} = \frac{n}{60^r} \tag{8-7}$$

The fraction on the right may be reduced to its lowest terms by cancellation of some factors. Since the base number 60 has the factorization into prime factors

$$60 = 2^2 \cdot 3 \cdot 5$$

it follows that in the reduced fraction a/b in (8–7), the denominator b can have only the prime factors 2, 3, and 5. This means that with suitable exponents α, β, and γ, we have

$$b = 2^\alpha \cdot 3^\beta \cdot 5^\gamma \tag{8-8}$$

Numbers of this type (8–8) may be called *regular* for the given base number 60. One sees conversely that if b is a regular number, the fraction (8–7) can be extended, and be written in the right-hand form as a fraction whose denominator is a power of 60, and so one finds a finite sexagesimal expansion.

The reader may verify that all entries in the table of inverses given above are regular, and also consider the question of finding the length of the expansion when b has a given prime factorization (8–8).

These discoveries in Babylonian mathematics also throw light on the history of early Greek science. The knowledge of Greek mathematics before Euclid has always been somewhat nebulous, and it has been difficult to understand the rapid rise from its primary stages, represented by Thales of Miletus (about 600 B.C.) and the Pythagoreans, to the beautiful system one finds developed at the time of Euclid (300 B.C.) or probably even earlier. It must now be assumed that the Greeks absorbed much more from the Babylonian storehouse of mathematical facts and methods than had hitherto been suspected. This, however, it should be explicitly stated, does not detract from the distinction of the Greeks for having created the concept of the systematic mathematical theory as we still understand and use it today, based upon axioms or fundamental assumptions and developed by logical deductions in its proofs. This achievement has been one of the most important in the history of human thought.

In the transition from Babylonia to Greece, mathematical knowledge changed its form. Greek mathematics is dominated

Fig. 8-2. Plimpton mathematical tablet 332. (*Columbia University Library. Courtesy Professor I. Mendelsohn.*)

by the geometric figure. This preference may in part be due to their feeling for beauty in lines and patterns, as shown in their decorative art and architecture, but a more compelling reason for the adoption of the geometric system was the logical consequence. The geometric lines were understandable and complete, while the numbers led to the logically incomprehensible, the unutterable concept of the irrational. Babylonian mathematics, on the other hand, was arithmetic and algebraic in character and expressed itself through numerical computations. Approximations were resorted to quite freely, thus obviating the necessity for the irrational perfectionism.

Judging from the advanced state of Babylonian mathematics

as revealed by the tablets, it seemed a reasonable conjecture that the Babylonians were in possession of the Pythagorean theorem. However, it was not until quite recently that a factual proof was found. In a new publication of cuneiform texts by Neugebauer and Sachs (1945), there is included a description of a clay tablet from the Plimpton Library at Columbia University, which bids fair to be one of the most crucial records in the history of mathematics. The tablet, catalogued as Plimpton 322, is composed in Old Babylonian script so that it must fall in the period from 1900 B.C. and 1600 B.C., at least a millenium before the Pythagoreans. Unfortunately, the tablet is broken and one section is missing, but there remain three complete columns of figures and part of a fourth which may be reconstructed (see Fig. 8–2). The reader may verify from the photographic reproduction that when we preserve the sexagesimal notation, the numbers in the three columns run as follows:

1, 59	2, 49	1
56, 7	1, 20, 25 [3, 12, 1]	2
1, 16, 41	1, 50, 49	3
3, 31, 49	5, 9, 1	4
1, 5	1, 37	5
5, 19	8, 1	6
38, 11	59, 1	7
13, 19	20, 49	8
8, 1 [9, 1]	12, 49	9
1, 22, 41	2, 16, 1	10
45	1, 15	11
27, 59	48, 49	12
2, 41 [7, 12, 1]	4, 49	13
29, 31	53, 49	14
56	1, 46 [53]	15

Clearly the last column only enumerates the lines. The first two columns are much more interesting. It is not difficult to verify that they form the hypotenuse and one leg of a Pythagorean triangle. When one squares the numbers in the middle column and subtracts from each of them the square of the corresponding number in the first column, one obtains a square number. There are, however, four exceptions to this rule, and in the preceding table the corrected figures have been given rather than the actual figures on the tablet, which have been put in brackets. The exception in line 2 is difficult to explain, while the number 9 instead of 8 in the ninth line must be a mere slip of the stylus. The number in line 13 is the square of the correct one and in line 15 half of the side occurred originally. It is of course somewhat unsatisfactory to be compelled to make four corrections in a table with 15 entry lines, but as we shall see, the fourth column gives a further check on the values in the other two columns, confirming again the corrected figures.

It is of interest to compute the missing column of the last side of the triangle and also to use our previous solution of the Pythagorean triangle given in (8-5) to determine the values of the numbers u and v that correspond to the solutions on the tablet. This information is given in the following table, which the reader may check:

b	c	a	u	v
119	169	120	12	5
3,367	4,825	3,456	64	27
4,601	6,649	4,800	75	32
12,709	18,541	13,500	125	54
65	97	72	9	4
319	481	360	20	9
2,291	3,541	2,700	54	25
799	1,249	960	32	15
481	769	600	25	12
4,961	8,161	6,480	81	40
45	75	60	2	1
1,679	2,929	2,400	48	25
161	289	240	15	8
1,771	3,229	2,700	50	27
56	106	90	9	5

All solutions are primitive except in line 11, where there is a common factor 15, and line 15, where there is a factor 2.

The question naturally arises whether the Babylonians were in possession of a method for solving the Pythagorean triangle corresponding to the general solution we have already established in (8–5). The answer must undoubtedly be in the affirmative, for many reasons. Of course one cannot hope to discover an explicit formula since no algebraic terminology existed at this time. In Babylonian mathematics, as in all early expositions, the reader was expected to infer the general rule from the examples given. Evidently, the large solutions of the Pythagorean problem found in the Plimpton Library tablet have not been obtained by guesswork; there are many much simpler solutions one would run across before these. The last leg of the triangle, computed in each case from the two given on the tablet, provides the key to the construction of the table. These numbers are all very simple in the sexagesimal system, as the reader may verify by rewriting them, and furthermore they are all regular sexagesimal numbers as we have defined this term, since they have only the prime divisors 2, 3, and 5. According to our solution (8–5), this side is determined by the formula $a = 2uv$ so that u and v are also regular sexagesimal numbers, as one sees by inspection of the table above. Thus it appears that the table on page 176 has been constructed by making a choice of small regular numbers for the parameters u and v.

This particular method had been used with a special idea in mind. The numbers representing the side a are all regular and occur in the tables of inverses; this fact points to their application in a division process, and indeed, the last, somewhat maculated, column on the tablet contains the value of the quotient c^2/a^2 for each triangle. If one denotes by α the angle in the right triangle opposing the side a (see Fig. 8–3), one has

$$\frac{c^2}{a^2} = \frac{1}{\sin^2 \alpha} = \operatorname{cosec}^2 \alpha$$

Another remarkable fact now becomes apparent. If one proceeds to compute the values of the quotients

$$\frac{c}{a} = \frac{1}{\sin \alpha} = \operatorname{cosec} \alpha$$

it is a consequence of the particular choice of the side a that this trigonometric function must have finite sexagesimal expansions. Furthermore, the values of cosec α form a very regular sequence with a decrease of almost exactly $1/60$ from one line to another,

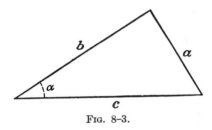

Fɪɢ. 8–3.

so that one would have a table of this trigonometric function constructed by means of right triangles with integral sides. Correspondingly, the angle decreases from $45°$ to $31°$, and it seems natural to believe that there existed companion tablets with similar values for the angles from $0°$ to $15°$ and from $16°$ to $30°$. How the Babylonians succeeded in finding values for c and a such that the quotient c/a decreases so evenly cannot be considered fully explained. It is evident, however, that at this early date the Babylonians not only had completely mastered the Pythagorean problem, but also had used it as the basis for the construction of trigonometric tables. One can only hope that future discoveries will produce further material, which will throw light upon this fascinating subject.

8–3. Diophantos of Alexandria. Greek mathematics at its height was preeminently geometric in character. However during the later Alexandrian period, when Greek science and philosophy as a whole was on the decline, and with it mathematics, the algebraic methods came more into the foreground. It is possible that this change may have been caused, at least to some degree, by outside influences from Babylon and perhaps even India.

During this period, Diophantos (perhaps A.D. 250), the most renowned proponent of Greek algebra, lived in Alexandria. Practically nothing is known about his life. There exists a collection of Greek problems in poetic form, the *Palatine Anthology*, which was compiled probably not over a century after Diophantos's death. It contains certain simple problems that can be solved by equations, some of them indeterminate, and among them one finds the following, containing all known personal information about Diophantos:

Here you see the tomb containing the remains of Diophantos, it is remarkable: artfully it tells the measures of his life. The sixth part of his life God granted him for his youth. After a twelfth more his cheeks were bearded. After an additional seventh he kindled the light of marriage, and in the fifth year he accepted a son. Elas, a dear but unfortunate child, half of his father he was and this was also the span a cruel fate granted it, and he consoled his grief in the remaining four years of his life. By this device of numbers, tell us the extent of his life.

If x is the age of Diophantos and if one interprets the poetic statement to mean that the son died at the age when he was half the father's ultimate age, the equation becomes

$$\frac{x}{6} + \frac{x}{12} + \frac{x}{7} + 5 + \frac{x}{2} + 4 = x$$

Thus $x = 84$ was his age.

The known titles of works of Diophantos are the *Arithmetics* in 13 books, the *Porisms*, and a treatise on polygonal numbers. All of them dealt with the properties of rational or integral numbers. Unfortunately, the *Porisms* have been lost and only a part of the *Polygonal Numbers* exists. Six or seven of the books of the *Arithmetics* have been preserved and there is some doubt whether the whole cycle was ever completed. In regard to the title it should be pointed out that Greek mathematicians used the term *arithmetic* in the sense of number theory, *i.e.*, the systematic investigation of the properties of numbers, while ordinary computations were classified as *logistics*.

In his mathematical presentation Diophantos uses stenographic abbreviations with special signs, often composed of the initial letters of the names of the concepts he wishes to designate; the unknown quantity, powers, and various operations therefore have fixed notations. This form of mathematical writing has been called *syncopated algebra*, and it must be considered an early step towards algebraic formalization and the creation of mathematical language.

The *Arithmetics* deal with topics on algebraic equations and more particularly with the solution of certain problems in which it is required to find rational numbers satisfying prescribed conditions. More than 130 problems of this latter type, of considerable variety, are discussed, and Diophantos shows great ingenuity in devising elegant methods for their solution. He is particularly adept at selecting the unknowns in such a manner that the algebraic conditions become easily manageable. We shall reproduce a few of his problems to illustrate the kind of problems he tackles. They should be prefaced by the general remark that negative or zero solutions are always excluded.

Problem 1 in Book II requires: To find two numbers such that their sum is in a given proportion to the sum of their squares.

In modern notation we would write

$$\frac{x^2 + y^2}{x + y} = p$$

where x and y are the numbers to be found and p the given proportion. This may be written

$$x^2 - xp + y^2 - yp = 0$$

and when it is considered to be a second-degree equation in x, one finds the solutions

$$x = \frac{p}{2} \pm \sqrt{\frac{p^2}{4} - y^2 + py} \tag{8-9}$$

Since the problem should be solved in rational numbers, the number under the square root sign must be a square. A typical device for expressing this condition is based upon the following observation: It is possible to express every number in the form

$$r = \frac{p}{2} + ty$$

by a suitable choice of the number t. The square number which occurs under the root sign in (8–9) can therefore be written

$$\frac{p^2}{4} - y^2 + py = \left(\frac{p}{2} + ty\right)^2 \tag{8–10}$$

When one performs the reduction, the terms that do not involve y drop out, and one factor y may be canceled. There remains a simple equation of the first degree, which gives

$$y = p\frac{1 - t}{1 + t^2} \tag{8–11}$$

For any rational value of t, the corresponding y in (8–11) makes the expression (8–10) a square

$$\frac{p^2}{4} - y^2 + py = \left[\frac{p}{2}\frac{(1 + 2t - t^2)}{1 + t^2}\right]^2$$

When this is substituted in (8–9), one finds for x two solutions

$$x = p\frac{1 + t}{1 + t^2}, \qquad x = p\frac{t(t - 1)}{1 + t^2} \tag{8–12}$$

The general solution of the problem is therefore given by (8–11) and (8–12) with rational t. Diophantos, of course, has no formulas, but he illustrates the methods for $p = 10$. His solution $x = 12$, $y = 6$ corresponds to $t = \frac{1}{3}$.

The majority of Diophantos's indeterminate problems require that one shall find certain sets of square or cube numbers with special properties, and the solution of the Pythagorean triangle often comes into play in his procedures. For instance, in Problem 22 in Book IV it is proposed: To find three numbers such that one is the mean proportional between the two others, and such that the difference between any two of them shall be a square number.

If $x > y > z$ are the three rational numbers to be found, it is required that the three differences

$$x - y = a^2, \qquad y - z = b^2, \qquad x - z = c^2$$

shall be square numbers. To satisfy the first two conditions one must have

$$x = y + a^2, \qquad z = y - b^2 \qquad (8\text{-}13)$$

with arbitrary a and b, but when these values are substituted in the third it reduces to

$$a^2 + b^2 = c^2$$

Therefore, the three numbers a, b, and c must form a rational Pythagorean triangle so that according to (8–6) we have

$$a = 2tr, \qquad b = (t^2 - 1)r, \qquad c = (t^2 + 1)r \qquad (8\text{-}14)$$

with rational values t and r. It remains to fulfill the condition that y be the mean proportional between x and z, that is,

$$y^2 = xz$$

According to (8–13) this may be written

$$y^2 = (y + a^2)(y - b^2)$$

and after reduction one finds

$$y = \frac{a^2 b^2}{a^2 - b^2}$$

From (8–13) follows further

$$x = \frac{a^4}{a^2 - b^2}, \qquad z = \frac{b^4}{a^2 - b^2}$$

To obtain the general solution we must substitute the values (8–14) for a, b, and c. Since the solutions are to be positive, one must choose the sides of the Pythagorean triangle such that $a > b$. The example on which Diophantos illustrates the procedure corresponds to the familiar case $b = 3$, $a = 4$, $c = 5$, and he finds

$$x = \tfrac{256}{7}, \qquad y = \tfrac{144}{7}, \qquad z = \tfrac{81}{7}$$

We shall give a few more examples of problems from the *Arithmetics* of Diophantos, on which the reader may try his skill.

Problem 29, Book II: Find two square numbers such that when one forms their product and adds either of the numbers to it, the result is a square.

Problem 7, Book III: Find three numbers such that their sum is a square and the sum of any two of them is a square.

Problem 9, Book III: Find three numbers in arithmetic series such that the sum of any two of them is a square.

Problem 15, Book III: Find three numbers such that the product of two of them minus the third is always a square.

Problem 11, Book IV: Find two numbers such that their sum is equal to the sum of their cubes.

Problem 18, Book VI: Find a Pythagorean triangle in which the length of the bisector of one of the acute angles is rational.

Quite appropriately, as a tribute to Diophantos's early contribution to the subject, algebraic problems in which one is required to find rational solutions are called *Diophantine problems*. In modern terminology this concept is usually narrowed somewhat to refer mainly to problems with integral solutions. As a consequence, even our previous linear indeterminate problems are commonly called *linear Diophantine problems* in spite of the fact that these problems were not discussed by Diophantos, probably because he considered them trivial. For instance any linear equation with integral coefficients may be solved rationally by giving arbitrary rational values to all unknowns except one and expressing the remaining one by the others.

It seems unlikely that the large collection of problems in the *Arithmetics* should be the creation of a single author, and some of them must have been gleaned from previous sources. However, any statements about the earlier history of Diophantine problems are entirely conjectural. It is possible that Greek algebra was further developed than our present sources indicate, and it is always an open guess that Babylonian mathematics embraced problems of this type.

8–4. Al-Karkhi and Leonardo Pisano. To the subsequent Greek mathematicians, all of them of very minor stature, and to the Arabs, Diophantos remained an outstanding name, almost synonymous with algebra itself. Among the Arab mathematicians, al-Karkhi of Bagdad, who died around A.D. 1030, was probably the most scholarly and original. Two of his works are known. One is the *Al-Kafi fil hisab* or *Essentials of Arithmetic* which is of an elementary character and gives the rules for computations. It is peculiar in that it avoids the use of Hindu numerals throughout, although they were at this time quite common in Bagdad. Among certain orthodox groups among the Arabs there seems to have been strong objection to the Hindu numbers, in many ways reminiscent of the opposition of the abacists in Europe to the same numbers a few centuries later.

Al-Karkhi's second work, the *Al-Fakhri*, is a much more important document in the history of mathematics. It derives its name from al-Karkhi's friend, the grand vizier in Bagdad at the time, to whom the treatise was dedicated. Al-Karkhi, in many ways, was the Arabic successor to Diophantos, even to the extent that the *Al-Fakhri* contains long sections that have been copied verbatim from the *Arithmetics*. The general plan of the two works is the same. Both contain basic algebraic theory with applications to equations and especially to problems that should be solved in rational numbers. Although al-Karkhi repeats many of Diophantos's problems, he develops the methods further and also introduces problems of quite different types. However, in terms of our present-day algebraic symbolism, most of them do not present great difficulties.

For instance, in Problem 1 in Section 5 in the *Al-Fakhri* it is requested to find such numbers that the sum of their cubes is a square number. This means that the equation

$$x^3 + y^3 = z^2$$

shall be solved in rational numbers. One can put

$$y = mx, \qquad z = nx$$

and by substitution and cancellation of x^2, one obtains

$$x = \frac{n^2}{1 + m^3}$$

where m and n may be arbitrary rational numbers. Al-Karkhi gives the special solution $x = 1$, $y = 2$, $z = 3$, a set of numbers that probably led to the problem's being put. The same method is clearly applicable to much more general rational problems, for instance,

$$ax^n + by^n = cz^{n-1}$$

and others for which al-Karkhi gives illustrations.

In several problems he asks for rational solutions to two simultaneous equations that may be included in the general type

$$x^3 + ax^2 = y^2, \qquad x^3 - bx^2 = z^2$$

where a and b are known integers. Again al-Karkhi puts

$$y = mx, \qquad z = nx$$

and from the two equations he derives

$$x = m^2 - a, \qquad x = n^2 + b$$

Since these two numbers must be the same, the condition

$$m^2 - n^2 = a + b$$

must be satisfied for m and n. Here one puts $m = n + t$ and obtains

$$2nt + t^2 = a + b$$

or

$$n = \frac{a + b - t^2}{2t}$$

From this value one finds m and in turn the general solution for x, y, and z in terms of the arbitrary rational number t. Al-Karkhi, of course, only gives examples, for instance

$$x^3 + 4x^2 = y^2, \qquad x^3 - 5x^2 = z^2$$

is found to have the solution $x = 21$.

In regard to mathematical knowledge the Middle Ages in Europe was a vacuous period with a single, brilliant star, Leonardo Pisano. Leonardo was a mathematician of great originality and creative power but also a direct successor to the Arabic mathematical school, much in the way al-Karkhi was heir to the knowledge and inspiration of Diophantos. Leonardo never mentions his sources, but he was educated in North Africa and traveled widely in the Eastern Mediterranean, and there can be no question that he was familiar with works of the leading Arabic mathematical writers. In Leonardo's main work, the *Liber abaci* (1202), one finds many problems that have been borrowed literally from the *Al-Fakhri*, and therefore sometimes originally from Diophantos; others have their source in al-Khowarizmi's *Al-Jabr wal-Muqabalah*.

Leonardo's fame was widespread, and true to the customs of the time, he was presented with challenge problems from near and far. Some of these were indeterminate problems. For instance, Master Theodorus, court philosopher to Emperor Frederic II proposed the problem of finding numbers x, y, and z such that all three expressions

$$x + y + z + x^2, \qquad x + y + z + x^2 + y^2,$$
$$x + y + z + x^2 + y^2 + z^2$$

become squares. This was a problem truly in the tradition of Diophantos, and Leonardo gives as one solution

$$x = \tfrac{16}{5}, \qquad y = \tfrac{48}{5}, \qquad z = \tfrac{144}{5}$$

The Emperor Frederic II was a sincere patron of learning and actively promoted the diffusion of Arabic knowledge in Europe. No wonder therefore that he took an interest in such an outstanding scholar as Leonardo. Probably in the year 1224, he was summoned to take part in a mathematical tournament, which was to be held in the presence of the emperor. The problems were formulated and presented by John of Palermo, another scholar belonging to the entourage of the emperor. Leonardo easily carried off the laurels by solving all problems in the most admirable manner.

One of the problems was the solution of a particular cubic equation, and after having shown that there could be no rational root to it, Leonardo proceeded to compute the real root in sexagesimal fractions with an accuracy that corresponds to 11 decimal places. A much simpler problem was the following, which we mention only because it belongs to a type that at the time enjoyed considerable popularity:

Three men own a share in a heap of coins; the first owns $\frac{1}{2}$, the second $\frac{1}{3}$, and the third $\frac{1}{6}$ of the total. The money is divided by having each man take an amount arbitrarily. The first man afterwards returns $\frac{1}{2}$ of the coins he has taken, the second $\frac{1}{3}$, and the third $\frac{1}{6}$. The money thus returned is divided into three equal shares, which are given to each man, and it turns out that now everyone has his proper part. How much money was there, and how much money did each obtain the first time? We leave the solution to the reader.

Here we are more interested in the following indeterminate problem proposed in the tournament: Find such a square number that when 5 is added or subtracted one also obtains squares. In mathematical symbols, one wishes to find a number x such that

$$x^2 + 5 = y^2, \qquad x^2 - 5 = z^2 \qquad (8\text{--}15)$$

Leonardo gives the solution $x = 3\frac{5}{12}$.

Again one cannot exclude the possibility that Leonardo may have been familiar with this kind of problem since it occurs in earlier Arab writings. However, in a treatise *Liber quadratorum* (1225) written shortly after the tournament, Leonardo returns to the problem and here his methods are entirely different from those used by Arab mathematicians.

Let us discuss the general problem of finding a square number such that when a given number h is added to it or subtracted from it one obtains other square numbers. This means that we must find a number x such that simultaneously

$$x^2 + h = a^2, \qquad x^2 - h = b^2 \qquad (8\text{--}16)$$

and determine for which h rational solutions x can exist. We shall first determine the solutions in integers and this depends, as we shall see again, on the Pythagorean triangle. When the second equation (8–16) is subtracted from the first, one has

$$2h = a^2 - b^2 = (a - b)(a + b) \qquad (8\text{--}17)$$

Since the left-hand side is even, a and b must both be odd or both even. Therefore, $a - b$ is even

$$a - b = 2k$$

and k must be a divisor of h, according to (8–17). It follows that

$$a + b = \frac{h}{k}$$

and by adding and subtracting the last two equations, one finds

$$a = \frac{h}{2k} + k, \qquad b = \frac{h}{2k} - k$$

When these two expressions are substituted in the original equations (8–16), there results

$$x^2 + h = \left(\frac{h}{2k} + k\right)^2 = \left(\frac{h}{2k}\right)^2 + h + k^2$$

$$x^2 - h = \left(\frac{h}{2k} - k\right)^2 = \left(\frac{h}{2k}\right)^2 - h + k^2$$

so that we have now only a single condition

$$x^2 = \left(\frac{h}{2k}\right)^2 + k^2$$

Therefore the three numbers

$$x, \quad \frac{h}{2k}, \quad k$$

form a Pythagorean triangle, and according to the solution we have obtained, we can write

$$x = t(m^2 + n^2), \qquad \frac{h}{2k} = t(m^2 - n^2), \qquad k = 2mnt$$

where t is some integer and the expressions in m and n define a primitive solution of the triangle. When we take the product of the last two expressions, we obtain as the general solution to (8–16)

$$x = t(m^2 + n^2), \qquad h = 4mn(m^2 - n^2)t^2 \qquad (8\text{–}18)$$

We shall make a slight reduction in this solution. Let us suppose that we have a solution x of (8–16), where x has the factor t and h at the same time the factor t^2

$$x = x_1 t, \qquad h = h_1 t^2$$

From the two equations

$$x_1{}^2 t^2 + h_1 t^2 = a^2, \qquad x_1{}^2 t^2 - h_1 t^2 = b^2 \qquad (8\text{–}19)$$

it follows that a and b have the factor t

$$a = a_1 t, \qquad b = b_1 t$$

After the factor t^2 has been canceled in (8–19), one has

$$x_1{}^2 + h_1 = a_1{}^2, \qquad x_1{}^2 - h_1 = b_1{}^2$$

en no further such reduction is possible, we shall say that we a *primitive solution*. When this reduction is applied to the solution becomes

$$x = m^2 + n^2, \qquad h = 4mn(m^2 - n^2) \qquad (8\text{–}20)$$

mbers m and n produce a Pythagorean triangle where ve no common factor. The hypotenuse x is then relatively the sides

$$m^2 - n^2, \qquad 2mn$$

hence x is relatively prime to h, and (8–20) must be a primitive solution to the problem in the sense just defined.

When one takes small values for m and n, one finds the following primitive solutions:

m	n	h	x
2	1	24	5
3	1	96	10
3	2	120	13
4	1	240	17
4	3	336	25

When the equations (8–16) in our problem are to have integral solutions, the number h must have the form we have derived in (8–18). However, to determine when a given number h can be represented in this manner is in itself a problem that is not easily settled in general. After Leonardo, many mathematicians returned to the problem and the permissible numbers even received a special name, a *congruum*. This nomenclature is now obsolete and must not be confused with the congruent numbers we shall study in the next chapter.

Leonardo established the following simple property: A congruum is divisible by 24. In the examples given above in the table this is immediately verified. To prove it in general, we recall that of the numbers m and n that give a primitive solution to a Pythagorean triangle, one is odd and the other is even. The product mn is therefore divisible by 2, and so h is divisible by 8 according to (8–20). It remains to show that h is divisible by 3. This is immediate when m or n is divisible by 3. When neither of them is divisible by 3, one can write

$$m = 3m_1 \pm 1, \qquad n = 3n_1 \pm 1$$

so that

$$m^2 - n^2 = 9m_1{}^2 - 9n_1{}^2 \pm 6m_1 \pm 6n_1$$

is divisible by 3.

It is now time to return to the original tournament problem (8–15) solved by Leonardo. Since in this case $h = 5$ is not divisible by 24, there can be no integral solutions. We therefore write x, a, and b as fractions with a common denominator

$$x = \frac{x_1}{d}, \qquad a = \frac{a_1}{d}, \qquad b = \frac{b_1}{d}$$

By substitution into (8–15) and clearing the fractions, one obtains

$$x_1{}^2 + 5d^2 = a_1{}^2, \qquad x_1{}^2 - 5d^2 = b_1{}^2 \qquad (8\text{–}21)$$

If there is to be any solution to (8–15) in rational numbers, it must be possible to find some integer d such that $5d^2$ is a congruum. Now in the condition

$$5d^2 = 4mn(m^2 - n^2)$$

it is a natural first attempt to make $m = 5$, and one must then satisfy

$$d^2 = 4n(5^2 - n^2)$$

By trying out the first few integers, one sees that $n = 4$ gives a square

$$d^2 = 4 \cdot 4(5^2 - 4^2) = 144 = 12^2$$

The values $n = 4$, $m = 5$, according to (8–20), result in the solution

$$x_1 = 5^2 + 4^2 = 41$$

for (8–21). Consequently

$$x = \frac{x_1}{d} = \frac{41}{12} = 3\frac{5}{12}$$

is a solution to Leonardo's problem, as he actually stated. To check the solution we have

$$\left(\frac{41}{12}\right)^2 + 5 = \left(\frac{49}{12}\right)^2, \qquad \left(\frac{41}{12}\right)^2 - 5 = \left(\frac{31}{12}\right)^2$$

There are no indications that John of Palermo was himself a prominent mathematician. On the other hand, when we look back upon the process that was required for the solution of his last problem, it is evident that it was not proposed haphazardly. It is difficult to escape the conclusion that it was a problem drawn from previous sources, and in that case most naturally from Arabic scholars whom he encountered on his native Sicily, which under Frederic II was one of the centers in the exchange of European and Oriental scholarship.

We have followed the development of Diophantine analysis through the Arabs to its transmittal to Europe through Leonardo Pisano. There exists, however, another branch of this field of number theory, which we must mention although we shall not pursue it in detail. The Hindus early became acquainted with the works of Diophantos, but their own number theory took an independent direction. We have already mentioned (Sec. 6-2) the method of the pulverizer, a variation of the algorism of Euclid, which gave the Hindus the solution of their linear indeterminate problems. But both in the *Brahma-Sputa-Siddhanta* by Brahmagupta and the *Bija-Ganita* by Bhaskara, one finds considerable space devoted to indeterminate problems of the type

$$cx^2 + 1 = y^2$$

and more generally

$$cx^2 + a = y^2$$

Not only are the rational solutions found; the integral solutions are also discussed. Later it has turned out that this kind of problem is of systematic importance for various mathematical questions, for instance, for continued fractions and number theory in quadratic fields, both subjects that are left out of this book with regret.

Problems.

The reader may try to find the rational solutions to the following equations or sets of equations, all taken from the *Al-Fakhri:*

1. $x^2 + 5 = y^2$
2. $x^2 - 10 = y^2$
3. $x^2 - 2x - 2 = y^2$
4. $10 - x^2 = y^2$

5. $10x - 8 - x^2 = y^2$

6. $x^2 + x = y^2$, $x^2 + 1 = z^2$

7. $x^3 + y^3 = z^2$

8. $x^2 y^3 = z^3$

9. $x^4 + y^4 = z^3$

10. $x^2 - y^3 = z^2$

11. $x^6 + 5y^2 = z^2$

8–5. From Diophantos to Fermat. Fermat represents a focal point in the history of number theory; in his work the radiating branches of earlier periods were united and their content recreated in a richer and more systematic form.

The path from Diophantos to Fermat, although long in time, is quite direct. During the Renaissance, at the rebirth of classical learning, numerous manuscripts of Greek mathematical works reached Western Europe. The general level of mathematics in Europe had been extremely low during the Middle Ages, so low that the Greek knowledge was a revelation whose true content at times was found to be intolerably hard to decipher. Among the works were copies of the writing of Diophantos, whose very name had until then been unknown, and they represented a severe challenge to the mathematicians of the sixteenth century.

The first reference to Diophantos in the Occident seems to have been made by Regiomontanus in 1462. He reported that he had discovered a manuscript of a certain Diophantos in the Vatican library and that he was interested in making a translation from the Greek, a task he never seems to have tackled.

The first printed edition and translation of Diophantos into Latin was published in Heidelberg in 1575, by the German professor Holzman, a name which he changed to the Greek form Xylander. To show the impact of Greek mathematics on the European scholars, let us reproduce a part of Xylander's foreword to his translation of the *Arithmetics*. He mentions that he had heard earlier of the existence of a Diophantos manuscript, but

. . . since no one had edited it, I gradually silenced my eagerness to know it, and buried myself in the mastery of the works of such arithmeticians as I could obtain, and in my own cogitations on the subject. Truth however compels me to offer with complete frankness the testimony which follows, however much to my disgrace. As for Cossica or Algebra, since, self-taught—except for the mute teachings of books, I had not only acquired command of the subject, but also had advanced to the point of adding, giving variety, and in places even of making corrections to what such great and devoted teachers as Christifer Rodolphus Silesius, Michaelus Stifelius, Cardanus, Nonius, and others had written about it, I fell into that mood of complacency, which Heraclitus called "The Holy malady";—in short I came to believe that in Arithmetic and Logistic "I was somebody". And in fact by not a few, and among them some true scholars, I was adjudged an Arithmetician beyond the common order. But when I first came upon the work of Diophantos, his method and reasoning so overwhelmed me that I scarcely knew whether to think of my former self with pity or with laughter. It has seemed worth while in this place to proclaim my former state of ignorance, and at the same time to give some hint of the work of Diophantos, which swept away from my befogged eyes the cloud of darkness which enveloped them. The treatment of surds I had mastered so well that I had even ventured to add to the inventiveness of others some things not inconsiderable, and these contributions in the field of arithmetic were accounted of no small importance in view of the difficulties of the subject, which had driven many from the whole subject of mathematics. But how much more brilliant a performance was it, in problems which seemed scarcely capable of solution even with the help of surds, and where surds bidden to till the soil of Arithmetic, true to their name, turned a deaf ear and failed, to carry the solution of the subtlest kind of problems to a point where surds are not invoked, and are not so much as even mentioned.

Xylander emphasizes with admiration how Diophantos is able to avoid irrational square roots or surds in his solutions. He puns, as one sees, on the *surd* or *deaf numbers*, a term we have taken over as a direct translation from Arab authors. This is in direct analogy with the "unspeakable" numbers of the Pythagoreans, and it is fully as satisfactory as our "irrational" roots, translated from the Greek αλογος or *without ratio*. Another ancient term in Xylander is the *Cossica* or *Rule of Coss*, which in early English texts most often appears as the *Cossick Art*, synonymous with algebra or equation theory. It refers to the common terminology of the time, of Italian origin, in which the unknown to be found in a problem is termed the *thing* or *cosa*.

Xylander's source manuscript was quite unsatisfactory and this is reflected in his translation.   Nevertheless, the book created much interest in problems of Diophantos's type.   In 1621 Bachet de Méziriac, whose acquaintance we have already made in connection with the linear indeterminate problems, published a new edition with notes and comments.   In some of these he sharply and somewhat ungratefully criticizes Xylander, whose earlier edition clearly had been of assistance to him.   However, Bachet's edition represents a great improvement.   Furthermore, it is very probable that it has the unique distinction of being the work that introduced Fermat to the problems of number theory.

Fermat possessed a well-worn copy of Bachet's Diophantos, which he also used as a notebook.   In the margin he jotted down several of his most important results as they occurred to him in connection with the related problems in Diophantos.   After Fermat's death the entire book, together with Fermat's notes, was published by his son Samuel (1670).

We shall discuss the content of a few of the various results indicated by Fermat in his marginal comments to Diophantos. Here one finds the result we have already mentioned in connection with the factorization of numbers, and which we prove in Chap. 11, namely, that every prime of the form $4n + 1$ can be represented as a sum of two integral squares in a single manner.   By means of the identity

$$(a^2 + b^2)(c^2 + d^2) = (ac \pm bd)^2 + (ad \mp bc)^2 \quad (8\text{--}22)$$

which was known to Leonardo Pisano and was used implicitly by Diophantos, one can represent the product of any two numbers that are sums of two squares as the sum of two other squares, and even in two different ways.   We have, for instance,

$$13 = 3^2 + 2^2, \qquad 37 = 6^2 + 1^2$$

and find by using (8–22)

$$13 \cdot 37 = 20^2 + 9^2 = 16^2 + 15^2$$

In the special case

$$2 = 1^2 + 1^2$$

one derives from (8–22)

$$2(a^2 + b^2) = (a + b)^2 + (b - a)^2$$

One concludes, therefore, that any product whose factors are 2 and primes of the form $4n + 1$, can be represented as the sum of two squares. Moreover, if one multiplies a sum of two squares by a square number

$$k^2(a^2 + b^2) = (ka)^2 + (kb)^2$$

the result is a sum of two squares. This leads to the criterion: If N is an integer and n^2 its largest square factor, so that

$$N = N_0 n^2$$

then N is the sum of two squares if the prime factors of N_0 are 2 and primes of the form $4n + 1$. Conversely, it may be shown that these are the only numbers that are the sum of two squares. Fermat also gives a formula for the number of such representations. We shall return to these questions in Chap. 11.

Examples.

1. The two numbers

$$56 = 7 \cdot 2^3, \qquad 99 = 3^2 \cdot 11$$

cannot be the sum of two squares since the prime factors 7 and 11 are not of the form $4n + 1$.

2. The number

$$1{,}105 = 5 \cdot 13 \cdot 17$$

can be represented as the sum of two squares. To find the representations we observe that

$$5 = 2^2 + 1^2, \qquad 13 = 3^2 + 2^2, \qquad 17 = 4^2 + 1^2$$

By application of the identity (8–22) one obtains

$$5 \cdot 13 = 8^2 + 1^2 = 7^2 + 4^2$$
$$5 \cdot 17 = 9^2 + 2^2 = 7^2 + 6^2$$
$$13 \cdot 17 = 14^2 + 5^2 = 11^2 + 10^2$$

and by a repeated application one finds the four representations

$$1,105 = 33^2 + 4^2 = 32^2 + 9^2 = 31^2 + 12^2 = 24^2 + 23^2$$

One verifies that there are no others.

Problem.

Which of the numbers 101, 234, 365, 1,947 can be written as the sum of two squares?

Not all numbers are the sum of two squares, as we just observed. Some, but not all, of the others, can be written as the sum of *three* squares. For instance, the prime 43 is not the sum of two squares, but one has

$$43 = 5^2 + 3^2 + 3^2$$

Similarly, not all integers are the sum of three squares, for instance, the prime 47 is not so representable, as one easily verifies, but it is the sum of *four* squares, even in two ways

$$47 = 6^2 + 3^2 + 1^2 + 1^2 = 5^2 + 3^2 + 3^2 + 2^2$$

Bachet made the conjecture that every positive integer can be written as the sum of at most four squares, and he verifies it for all numbers up to 120. Fermat states in one of the Diophantos notes that he has a proof for this theorem. In a letter to the French mathematician Roberval he returns to the difficulties he had to overcome to find a proof and explains that he had finally succeeded through the use of his favorite method of *infinite descent,* a procedure he also had used to derive the results regarding the representation of numbers as the sum of two squares. He continues: "I confess openly that in the theory of numbers I have found nothing which I have enjoyed more than the proof of this theorem and I should be pleased if you would attempt to find it, even if it were only to let me know whether I value my discovery higher than it deserves."

There seems to be little reason to doubt that Fermat was in possession of a proof according to the indications he has given. That the problem was difficult can be judged from the fact that even the resourceful Euler in vain pitted his ingenuity against it, and not until 1770 did the French mathematician J. L. Lagrange, the successor of Euler at the Academy in Berlin and later a friend of Napoleon, publish the first proof. As so often happens, the completion of one problem gives birth to another.

In the same year Edward Waring (1734–1798), professor in Cambridge and both scientifically and personally one of England's most peculiar mathematicians, published his *Meditationes algebraicae*. In this work one finds several announcements and conjectures on the theory of numbers, among them the fact that every number can be represented as the sum of a limited number of cubes, fourth, or higher powers. This Waring's problem has occupied the mathematicians intensely. That such representations exist was proved by the German mathematician D. Hilbert in 1909. Essential information regarding the number of powers that are required in each case has been given by various mathematicians; among the most important results, one should mention particularly those of the English mathematicians G. H. Hardy and J. E. Littlewood, Vinogradoff (Russian), and L. E. Dickson of the University of Chicago.

8-6. The method of infinite descent. Fermat's method of the *descente infinie* is illustrated by his comments on Problem 26 in Book VI in Diophantos's *Arithmetics*. These remarks are interesting in several ways. He begins by stating: *"The area of a rational right triangle cannot be a square number.* The proof of this theorem I have reached only after elaborate and ardent study. I reproduce the proof here, since this kind of demonstration will make possible wonderful progress in number theory." Then follows a fairly complex indication of the proof and it is remarkable that in the long statement he uses no mathematical symbolism whatever, giving all terms in longhand words. Towards the end he breaks off with the statement: "The margin is insufficient to give all details of the proof."

We shall give the proof in ordinary algebraic symbols, but we first reduce the problem to integers by the following observations. When the area of a rational triangle is a square number and each side is multiplied or divided by a factor, the area is multiplied or divided by the square of this factor and it remains a square number. One can therefore clear the fractions in the rational sides to make them integral, and if they now should have any common divisor, it may be canceled. It follows that it is sufficient to show that the area of an integral, primitive Pythagorean triangle cannot be a square number.

The proof of this theorem is a bit long, but each step, as will be seen, is quite simple. The sides of a primitive Pythagorean triangle

$$a = 2mn, \qquad b = m^2 - n^2, \qquad c = m^2 + n^2 \qquad (8\text{-}23)$$

we have already found. The area of the triangle is

$$A = \tfrac{1}{2}ab = mn(m^2 - n^2) \qquad (8\text{-}24)$$

Since this integer shall be a square, one must have

$$mn(m - n)(m + n) = t^2 \qquad (8\text{-}25)$$

In a primitive triangle the numbers m and n are relatively prime, one even and the other odd, and one concludes, therefore, that among the four numbers

$$m, \quad n, \quad m - n, \quad m + n$$

any two are relatively prime. Since their product is a square, according to (8-25), each one of them is a square

$$m = u^2, \qquad n = v^2, \qquad u^2 - v^2 = p^2, \qquad u^2 + v^2 = q^2 \qquad (8\text{-}26)$$

where all four numbers u, v, p, q, also must be relatively prime in pairs.

By adding and subtracting the last two equations in (8-26), one finds

$$2u^2 = p^2 + q^2, \qquad 2v^2 = q^2 - p^2 = (q - p)(q + p) \qquad (8\text{-}27)$$

Since one of the numbers m and n is odd and the other even, u and v must have the same property, so that according to (8-26) p and q must both be odd. This shows further that $q - p$ and $q + p$ are both even so that the second equality in (8-27) yields that v is even

$$v = 2v_1$$

We put this into the second equation in (8–27) and find

$$2v_1{}^2 = \frac{q - p}{2} \cdot \frac{q + p}{2} \tag{8-28}$$

Here the two factors $\frac{q - p}{2}$ and $\frac{q + p}{2}$ on the right are relatively prime, because a common factor would divide their sum q and their difference p, but p and q are relatively prime.

In (8–28) we have two alternatives, depending on which of the factors on the right is even. Let us suppose that $\frac{q - p}{2}$ is even. Then, besides the factor 2 it must have some factor in common with $v_1{}^2$, while $\frac{q + p}{2}$ must be equal to the remaining factor of $v_1{}^2$, so that we can write

$$\frac{q - p}{2} = 2k^2, \qquad \frac{q + p}{2} = l^2 \tag{8-29}$$

where

$$v_1{}^2 = k^2 l^2 \tag{8-30}$$

The other alternative is that $\frac{q + p}{2}$ is even, and in this case one obtains similarly

$$\frac{q - p}{2} = l^2, \qquad \frac{q + p}{2} = 2k^2 \tag{8-31}$$

where (8–30) still holds.

From (8–29) one finds

$$p = l^2 - 2k^2, \qquad q = l^2 + 2k^2$$

and, in the alternative case (8–31),

$$p = 2k^2 - l^2, \qquad q = 2k^2 + l^2$$

When these values for p and q are substituted in the first equation (8–27), one finds in both cases

$$u^2 = (l^2)^2 + (2k^2)^2 \tag{8-32}$$

We have now completed the steps preparatory to the use of the infinite-descent argument. Our starting point was the integral Pythagorean triangle (8–23) with an area (8–24) which was supposed to be an integral square number. From this we have, according to (8–32), derived a new triangle of the same kind with the sides

$$l^2, \quad 2k^2, \quad u$$

The area of the new triangle is found to be

$$A_1 = \tfrac{1}{2} \cdot l^2 \cdot 2k^2 = l^2k^2 = v_1{}^2 = \left(\frac{v}{2}\right)^2 = \frac{n}{4}$$

so that it is also an integral square number, which clearly is smaller than the area of the original triangle. From this second triangle one could derive a third, a fourth, and so on, with the same properties and steadily decreasing integral areas. This, however, clearly involves a contradiction since the area is always an integer $\geqq 1$. Our initial assumption that there existed some Pythagorean triangle with a square number for its area is therefore inacceptable, and the theorem is proved.

In general, the method of infinite descent may be stated in the following form: it is assumed that a problem can be solved in positive integers and one derives from this a new solution in smaller numbers; since positive integers cannot be decreased indefinitely, one arrives at a contradictory situation so that the assumption that the problem had a solution is impossible.

There are various consequences of the result that the area of a Pythagorean triangle cannot be square. Let us return for a moment to the concept of a congruum, which was introduced in connection with the problem of Leonardo. The expression (8–20) for a congruum was

$$h = 4mn(m^2 - n^2)$$

and when one compares it with the expression (8–24) for the area of a Pythagorean triangle, one sees that the congruum is four times the area of the triangle defined by m and n. We can state therefore: A congruum cannot be a square number. Leonardo was aware of this result in his *Liber quadratorum*, but he did not possess any satisfactory proof for it.

Another consequence is: There are no fourth powers whose difference is a square; *i.e.*, the equation

$$x^4 - y^4 = z^2 \tag{8-33}$$

has no solution in integers x, y, and z different from zero.

If one could find two such integers x and y fulfilling the equation (8–33), the Pythagorean triangle defined by $m = x^2$, $n = y^2$ would, according to (8–24), have the area

$$x^2y^2(x^4 - y^4) = x^2y^2z^2$$

which is a square.

8–7. Fermat's last theorem. We now come to the most famous of Fermat's remarks in his copy of Diophantos. In Problem 8 in Book II Diophantos propounds: To decompose a given square number into the sum of two squares.

To use a general notation, let a^2 be the given square for which one wants to find x and y such that

$$a^2 = x^2 + y^2 \tag{8-34}$$

As usual, Diophantos asks for rational solutions. For a suitably chosen number m, one can then write

$$y = mx - a$$

When this is substituted into (8–34), one can cancel a factor x and find

$$x = \frac{2am}{m^2 + 1}$$

Here m may be any rational number. Diophantos must proceed only by illustrating the method on an example. He chooses $a = 4$ and takes a solution that corresponds to $m = 2$ in our formula, giving

$$x = \tfrac{16}{5}, \qquad y = \tfrac{12}{5}$$

One verifies that

$$(\tfrac{16}{5})^2 + (\tfrac{12}{5})^2 = 4^2$$

This problem to us is quite straightforward, but it was not always so. In the oldest preserved Diophantos manuscript, copied in the thirteenth century, we find at this point the following heartfelt remark by the writer: "Thy soul, Diophantos, to Satanas, for the difficulty of thy problems and this one in particular."

Fermat's comments in connection with this problem are, as one should expect, considerably more constructive and of much

Fɪɢ. 8–4. Pierre de Fermat (1608–1665).

greater consequence: "However, it is impossible to write a cube as the sum of two cubes, a fourth power as the sum of two fourth powers and in general any power beyond the second as the sum of two similar powers. For this I have discovered a truly wonderful proof, but the margin is too small to contain it."

This is the famous *Fermat's theorem*, sometimes called *Fermat's last theorem*, on which the most prominent mathematicians have tried their skill ever since its announcement three hundred years ago. In algebraic language, it requires that it shall be shown that the Diophantine equation

$$x^n + y^n = z^n \qquad (8\text{–}35)$$

has no solution in integers x, y, and z, all different from zero, when $n \geqq 3$.

For one case of the theorem Fermat obviously had a proof. It follows from our previous results that the equation

$$x^4 + y^4 = z^4 \tag{8-36}$$

cannot have any integral solutions different from zero. One can write this equation in the form

$$z^4 - y^4 = (x^2)^2$$

and since we have shown that the difference between two fourth powers cannot be a square, (8–36) is also impossible in integers. This result also goes a little further. If the exponent n in (8–35) is divisible by 4, one can write $n = 4m$, and Fermat's equation takes the form

$$(x^m)^4 + (y^m)^4 = (z^m)^4$$

and this equation is impossible as we have just shown.

By a similar remark one can reduce the general case to the case where the exponent in (8–35) is an odd prime. Let us suppose that $n = pm$ where $p > 2$ is a prime. Then Fermat's equation may be written

$$(x^m)^p + (y^m)^p = (z^m)^p$$

so that it is sufficient to prove the equation impossible for prime exponents $p > 2$.

The question whether Fermat possessed a demonstration of his last problem will in all likelihood forever remain an enigma. Fermat undoubtedly had one of the most powerful minds ever applied to investigate the laws of numbers, and from his indications there is every reason to believe that he was able to prove the various other assertions that he included in the Diophantos notes. The remark that the margin was too small may perhaps sound a bit like an excuse, but it was an observation he had to make also in other instances. On the other hand, he may have made a mistake, as in another case, where the conjecture, which he repeated in several letters, that all Fermat numbers were primes proved incorrect. Mathematicians occasionally may argue the point: the consensus seems to be that in view of the numerous investigations of the problem for three centuries from every conceivable angle, by first- and second-rate mathematicians, by amateurs and dilettanti, it is very unlikely that there should exist a proof based on any methods one can reasonably assume Fermat could have mastered. Such methods would undoubtedly have great consequences in other problems of number theory, but Fermat mentions them nowhere. Like so many of the other mathematicians who later worked on the problem, including Kummer whose results were the most incisive of all, he may have fallen into one of the many pitfalls of insufficient reasoning that have beset the investigations on the problem.

Fermat's problem has remained remarkably active throughout its history, and results and research on it still appear frequently in the mathematical journals. It must be admitted frankly that if the specific result implied in the theorem were obtained, it would probably have little systematic significance for the general progress of mathematics. However, the theorem has been extremely important as a goal and a constant source of new efforts. Some of the new methods it has inspired have proved to be basic not only for number theory but also for many other branches of mathematics.

As we mentioned, Fermat gave a proof of his theorem when $n = 4$. The case $n = 3$ he presented repeatedly as a challenge problem to French and English mathematicians, and it seems unlikely that he should propose a problem to which he could not himself give an answer, if requested. The first proof for the cubic case was published by Euler in a French translation of his *Algebra*. The case $n = 5$ was proved independently about 1825 by the German mathematician Lejeune-Dirichlet and the French Legendre, and the case $n = 7$ in 1839 by Lamé.

The most significant advance in the investigations of the problem was made by the German mathematician E. Kummer (1810–1893). He extended the domain of number theory to include not only the rational numbers but also the *algebraic numbers, i.e.,* numbers satisfying algebraic equations with rational coefficients.

In 1843 Kummer submitted to Lejeune-Dirichlet a manuscript containing a purported proof of Fermat's theorem based on algebraic numbers. Dirichlet, who had made similar attempts himself, immediately picked out the error in the reasoning: in the domain of algebraic numbers the fundamental theorem no longer holds that every number is representable essentially in one way as a product of prime factors. This failure caused Kummer to attack the problem with redoubled vigor, and a few years later he succeeded in finding a substitute for the theorem of the unique factorization in the theory of *ideals*, a theory that later has gained importance in almost all parts of mathematics.

By means of the ideals, Kummer was able to derive very general conditions for the insolubility of Fermat's theorem. Practically all important progress in this field in the last century has been made along the lines suggested by the theory of Kummer. Numerous criteria have been developed by means of which Fermat's equation has been proved impossible for all exponents at least up to $n = 600$.

A curious twist was added to the history of Fermat's problem in 1908 when the German mathematician P. Wolfskehl, who had made a few contributions related to the subject, bequeathed 100,000 marks to the Academy of Science in Göttingen for a prize to be awarded for the first complete proof of Fermat's last theorem. The prize probably added little or nothing to the interest of the mathematicians in the problem, but an immediate consequence was a deluge of alleged proofs by laymen eager to gain money and glory. This interest of the dilettanti in the problem has since never quite ceased, and Fermat's problem has without question the distinction of being the mathematical problem for which the greatest number of incorrect proofs have been published. (See Supplement.)

Bibliography

BONCONPAGNI, B.: *Scritti di Leonardo Pisano, matematico del secolo decimo terza*, Rome, 1857–1862.

CARMICHAEL, R. D.: *Diophantine Analysis*, John Wiley & Sons, Inc., New York, 1915.

COLEBROOK, H. T.: *Algebra with Arithmetic and Mensuration from the Sanscrit of Brahmegupta and Bhascara,* London, 1817.

FERMAT, P.: *Oeuvres,* publiées par les soins de M. M. Paul Tannery et Charles Henry, 4 vols., Paris, 1891–1912.

HEATH, T. L.: *Diophantos of Alexandria; A Study in the History of Greek Algebra,* second edition, with a supplement containing an account of Fermat's theorems and problems connected with Diophantine problems by Euler, Cambridge University Press, London, 1910.

MORDELL, L. J.: *Three Lectures on Fermat's Last Theorem,* Cambridge University Press, London, 1921.

NEUGEBAUER, O.: *Mathematische Keilschrifttexte,* 3 vols., Verlag Julius Springer, Berlin, 1935–1937.

―――― and A. SACHS: *Mathematical Cuneiform Texts,* American Oriental Series, Vol. 29, New Haven, 1945.

THUREAU-DANGIN, F.: *Textes mathématiques babyloniens,* Leiden, 1938.

VANDIVER, H. S.: "Fermat's Last Theorem, Its History, and the Nature of the Known Results Concerning It," *American Mathematical Monthly,* Vol. 53, 555–578 (1946).

WOEPKE, F.: *Extrait du Fakhri,* Paris, 1853.

CHAPTER 9

CONGRUENCES

9–1. The Disquisitiones arithmeticae. Who were the greatest mathematicians of all times? If one should put this question to a gathering of mathematicians, there would of course be disagreement, but a considerable number would undoubtedly state as their choices: Archimedes, Newton, and Gauss. Among these, Gauss is the only one whose work made an essential contribution to the theory of numbers. One could go further and state that while Fermat was the father of number theory as a systematic science, Gauss inspired the modern phase of the subject. His most important work on the properties of numbers is the *Disquisitiones arithmeticae*, which appeared in 1801 when he was twenty-four years of age.

Carl Friedrich Gauss (1777–1855) was the son of a bricklayer who on the whole was quite opposed to the idea of an advanced education for the boy. The young Gauss was, however, a precocious child whose ability so overwhelmed his teachers that as a fourteen-year-old boy he was presented to Carl Wilhelm Ferdinand, the Duke of Brunswick. The duke financed his education and granted him a small pension on which he lived until the tragic death of the duke in 1806, on the flight from Napoleon's armies. The next year Gauss was appointed director of the university observatory in Göttingen. Here he lived until his death, secluded and reserved, caring little for students and pupils, indifferent to honors, but bringing forth from time to time some masterpiece of mathematical creation. His contemporaries looked up to him with awe and universally acclaimed him the *princeps mathe-*

maticorum. His contributions covered practically all fields of mathematics, pure and applied, including mechanics, astromony, physics, geodesy, and statistics.

The *Disquisitiones arithemeticae* has often been pronounced the greatest among his many great works, both in results and in the depth of its new ideas. Many problems, some of them previously

Fig. 9-1. Carl Friedrich Gauss (1777–1855).

attacked in vain by prominent mathematicians, here received their solution for the first time. In the opening sections Gauss introduces a new calculus, the *theory of congruences*, that almost immediately gained general acceptance and ever since has put its stamp on all terminology in number theory. The subsequent chapters will be applied to the discussion of various aspects of this theory.

In a devoted statement Gauss dedicates the *Disquisitiones* to his patron, the Duke of Brunswick and Lunebourg, praising him particularly because he had been willing to lend his support also to those parts of science "which appear most abstract and with less application to ordinary usefulness, because in the depth of your wisdom, able to profit by all which tends to the happiness and prosperity of society, you have felt the intimate and necessary liaison which unites all sciences." In the introduction Gauss mentions earlier investigations in the theory of numbers, particularly those by Euclid and Diophantos as well as those by Fermat, Euler, Lagrange, and Legendre. He relates that he began his research in the theory of numbers when he was eighteen years old and that he had been so attracted to these questions that a considerable part of the *Disquisitiones* had been completed before he became familiar with the results of other mathematicians. But "reading the works of these men of genius I was not late in recognizing that I had employed the greater part of my meditations on things known for a long time; but animated by a new ardor in following their steps, I exerted myself to advance further the cultivation of number theory." In the final presentation he included many of his earlier and previously known results to give a systematic view of the whole field.

9–2. The properties of congruences. Gauss introduces his congruences through the following definition: Two integers a and b shall be said to be *congruent for the modulus m* when their difference $a - b$ is divisible by the integer m. This he expresses in the symbolic statement

$$a \equiv b \ (\text{mod } m) \tag{9–1}$$

When a and b are not congruent, they are called *incongruent for the modulus m* and this is written

$$a \not\equiv b \ (\text{mod } m)$$

These terms, as one sees, are derived from Latin, *congruent* meaning *agreeing* or *corresponding* while *modulus* signifies *little measure*. The latter term is often shortened to *modul*.

Let us illustrate the definitions by a few examples. One has for instance

$$26 \equiv 16 \ (\text{mod } 5)$$

since the difference $26 - 16 = 10$ is divisible by 5; also

$$12 \equiv 39 \ (\text{mod } 9)$$

since $12 - 39 = -27$ is divisible by 9, while

$$3 \not\equiv 11 \pmod 7$$

because $3 - 11 = -8$ is not divisible by 7.   Gauss uses the examples

$$16 \equiv -9 \pmod 5$$
$$-7 \equiv 15 \pmod{11}$$
$$-7 \not\equiv 15 \pmod 3$$

One can state the congruence (9–1) slightly differently by saying that b is congruent to a when it differs from a by a multiple of m

$$b = a + km \tag{9-2}$$

There are certain basic properties of congruences, which we shall enumerate.   The first is

1. *Determination.*   For any pair of integers a and b one has one or the other of the alternatives

$$a \equiv b \pmod m, \quad a \not\equiv b \pmod m$$

In other words, either the difference $a - b$ is divisible by m or it is not.   The second property is equally trivial:

2. *Reflexivity.*   One has

$$a \equiv a \pmod m$$

This states only that $a - a = 0$ is a multiple $0 \cdot m$ of any number m.

3. *Symmetry.*   When

$$a \equiv b \pmod m$$

then one also has

$$b \equiv a \pmod m$$

This is clear since when the difference $a - b$ is divisible by m so is $b - a$.   The last property of this kind is

4. *Transitivity.*   When

$$a \equiv b \pmod m, \quad b \equiv c \pmod m$$

then

$$a \equiv c \pmod m$$

To prove it we need only observe that

$$a - c = (a - b) + (b - c)$$

is divisible by m according to the first two congruences.

These four properties 1–4 show that the congruences for some given modul define a relation between any two numbers of a type that in mathematics is called an *equivalence relation*. The best-known example of such a relation is the ordinary equality

$$a = b$$

It may be of interest to observe that the equality may itself be considered to be a congruence, namely for the modulus 0, since according to (9–2) the congruence

$$a \equiv b \pmod 0$$

signifies that $a = b$. This artificial terminology is not in use.

There is, however, another relation that may be expressed conveniently by means of congruences. As one sees immediately from the definition of a congruence, the fact that a number a is *divisible* by a number m may be stated

$$a \equiv 0 \pmod m$$

For example, one has

$$6 \equiv 0 \pmod 2, \qquad 35 \equiv 0 \pmod 5, \qquad 13 \not\equiv 0 \pmod 7$$

The even numbers n are characterized by

$$n \equiv 0 \pmod 2$$

Problems.

Verify the congruences

1. $40 \equiv 13 \pmod 9$

2. $7 \not\equiv 99 \pmod{13}$

3. $3^4 \equiv 1 \pmod 5$

4. $11 \equiv 23 \pmod{12}$

5. $132 \equiv 0 \pmod{11}$

6. $7^2 \equiv 1 \pmod 8$

9–3. Residue systems. When an integer a is divided by another m, one has

$$a = km + r \tag{9–3}$$

where the remainder r is some positive integer less than m. Thus for any number a there exists a congruence

$$a \equiv r \pmod{m}$$

where r is a unique one among the numbers

$$0, 1, 2, \ldots, m-1 \qquad (9\text{-}4)$$

For this reason the set (9-4) is called a *complete residue system (mod m)*. One has for instance

$$35 \equiv 2 \pmod{11}, \qquad -11 \equiv 5 \pmod{8}$$

On the other hand, all numbers a that are congruent to a given remainder r in (9-4) will be of the form (9-3), where k is an arbitrary integer. Since these are the numbers that correspond to the same remainder r when divided by m, we say that they form a *residue class (mod m)*. There are m residue classes (mod m). For a given remainder r the residue class to which it belongs consists of the numbers

$$r, \quad r \pm m, \quad r \pm 2m, \quad \cdots$$

According to our definition the congruence

$$a \equiv b \pmod{m}$$

signifies that the numbers a and b differ by a multiple of m; consequently the congruence can also be expressed in the terms that a and b belong to the same residue class (mod m).

NOTE: We have previously (Sec. 7-3) introduced the term *modul* as a set of numbers closed with respect to addition and subtraction. This is a somewhat different concept from the congruence modul just defined; in the following we shall use the term only for congruences so that no confusion can arise. The two concepts are, however, closely related. We showed in theorem 7-6 that a modul as a set of integers consisted of all multiples

$$0, \pm m, \pm 2m, \cdots$$

of an integer m, so that this set is the *zero residue class* (mod m), *i.e.*, the set of all numbers a for which the congruence

$$a \equiv 0 \pmod{m}$$

is fulfilled.

Examples.

1. For the modulus $m = 2$ there are two remainders, **0 and 1**, and the corresponding residue classes are

$$\cdots, \quad -4, \quad -2, \quad 0, \quad 2, \quad 4, \quad \cdots$$
$$\cdots, \quad -3, \quad -1, \quad 1, \quad 3, \quad 5, \quad \cdots$$

consisting respectively of the even and odd numbers.

2. When $m = 3$ there are three residue classes

$$\cdots, \quad -6, \quad -3, \quad 0, \quad 3, \quad 6, \quad \cdots$$
$$\cdots, \quad -5, \quad -2, \quad 1, \quad 4, \quad 7, \quad \cdots$$
$$\cdots, \quad -4, \quad -1, \quad 2, \quad 5, \quad 8, \quad \cdots$$

3. Prove that all numbers in a residue class have the same g.c.d. with the modulus m.

There are many other residue systems such that every number is congruent (mod m) to a single one among them. We may recall, for instance, that in the division process we sometimes found it convenient to use least absolute remainders. In general, we shall say that m numbers

$$a_1, \ a_2, \ldots, \ a_m \tag{9-5}$$

form a *complete system of residues* if every number is congruent to some a_i. One sees that to obtain such a system one must pick one a_i from each of the m residue classes. To examine whether the numbers (9-5) form a complete residue system, one can verify that they are congruent to the numbers (9-4) in some order. For instance, the numbers

$$32, \ -1, \ 8, \ 20, \ 11$$

form a complete system of residues (mod 5) since one has the congruences

$$32 \equiv 2, \qquad -1 \equiv 4, \qquad 8 \equiv 3,$$
$$20 \equiv 0, \qquad 11 \equiv 1 \qquad (\bmod 5)$$

Another way of determining that the m numbers (9–5) form a complete system of residues would be to show that no two of them are congruent

$$a_i \not\equiv a_j \ (\bmod m)$$

since in this case they would all belong to different residue classes.

Problems.

1. Show that the numbers

$$-3, \quad 14, \quad 3, \quad 12, \quad 37, \quad 50, \quad -1$$

form a complete residue system (mod 7).

2. Do the numbers

$$5, \quad 12, \quad -3, \quad -4, \quad 9, \quad 22$$

form a complete residue system (mod 6)?

9–4. Operations with congruences. We began by emphasizing that some of the basic properties of congruences are the same as those of ordinary equality. We shall now pursue this analogy further and establish that one can operate with congruences according to rules that in many ways resemble those used in combining equations. A little later on we shall show that several important applications of congruences depend on this fact.

This first property we mention is:

THEOREM 9–1. Congruences for the same modul may be added and subtracted. If

$$a \equiv b; \qquad c \equiv d \qquad (\bmod m) \qquad (9\text{-}6)$$

then

$$a + c \equiv b + d; \qquad a - c \equiv b - d \qquad (\bmod m)$$

To prove, for instance, the first one of these congruences, it is sufficient to observe that the difference

$$a + c - (b + d) = (a - b) + (c - d)$$

is divisible by m according to the two given congruences (9–6). As an example, we may take

$$5 \equiv 32; \quad 11 \equiv -7 \quad (\text{mod } 9)$$

By addition and subtraction one finds the new congruences

$$16 \equiv 25; \quad 6 \equiv -39 \quad (\text{mod } 9)$$

which are also seen to be correct.

By repeated application of the addition rule, it follows that one can add an arbitrary set of congruences for the same modulus. For instance, from the three congruences

$$47 \equiv -5 \ (\text{mod } 13)$$

$$11 \equiv 37 \ (\text{mod } 13)$$

$$1 \equiv -25 \ (\text{mod } 13)$$

one obtains by addition of the numbers on both sides

$$59 \equiv 7 \ (\text{mod } 13)$$

Another application of the addition theorem results in:

THEOREM 9–2. A congruence may be multiplied by an arbitrary integer. From

$$a \equiv b \ (\text{mod } m)$$

it follows that

$$ka \equiv kb \ (\text{mod } m)$$

Clearly the new congruence has been obtained by adding the given congruence to itself k times. For instance, from

$$3 \equiv 7 \ (\text{mod } 4)$$

one concludes by multiplication with 5 that

$$15 \equiv 35 \ (\text{mod } 4)$$

The next result is:

THEOREM 9–3. Two congruences may be multiplied together. From the congruences in (9–6) one obtains

$$ac \equiv bd \ (\text{mod } m) \qquad (9\text{–}7)$$

This may be derived in various ways; one can, for instance, multiply the first congruence (9–6) by c and the second by b so that

$$ac \equiv bc \equiv bd \pmod{m}$$

One can also express the difference between the two sides in the congruence (9–7) in the form

$$ac - bd = (a - b)c + b(c - d)$$

showing that it is divisible by m. To illustrate, the multiplication of the two congruences

$$3 \equiv 14, \qquad 9 \equiv -2 \qquad \pmod{11}$$

gives

$$27 \equiv -28 \pmod{11}$$

Again the multiplication rule may be applied to several congruences. In particular, a congruence may be multiplied by itself any number of times so that

$$a \equiv b \pmod{m}$$

implies

$$a^n \equiv b^n \pmod{m}$$

for any exponent n.

Since any of the operations of addition, subtraction, and multiplication when applied to congruent numbers will give congruent results, we conclude that any algebraic expression constructed by repeated use of these operations will give congruent results when congruent values are substituted. For instance, since

$$-2 \equiv 3 \pmod{5}$$

the polynomial

$$f(x) = x^3 - 8x + 6$$

must give congruent results when -2 and 3 are substituted. One finds actually

$$f(-2) = 14 \equiv 9 = f(3) \pmod{5}$$

The same would hold if one took two polynomials in which the corresponding coefficients were congruent. For instance, the polynomials

$$f(x) = x^3 - 8x + 6, \qquad g(x) = 4x^3 - 3x^2 - 2x - 3$$

have congruent coefficients (mod 3), namely,

$$1 \equiv 4, \qquad 0 \equiv -3, \qquad -8 \equiv -2, \qquad 6 \equiv -3 \qquad \text{(mod 3)}$$

Thus the two values $x = -2$ and $x = 1$, which are congruent (mod 3), must give congruent values when substituted in $f(x)$ and $g(x)$, respectively. One sees that

$$f(-2) = 14 \equiv -4 = g(1) \text{ (mod 3)}$$

Analogous results must hold if one takes expressions with several variables.

These rules for the computation with congruences are, as we have seen, quite simple and analogous to those for equations. Nevertheless, the reader who makes his beginning steps with this somewhat unfamiliar and strange calculus will need a little time and several examples to gain the necessary confidence in the method. After some experience it will become clear how much the notion of congruences facilitates certain kinds of considerations in number theory.

Examples.

1. Let us determine the smallest positive remainder (mod 17) of the number 37 when raised to the thirteenth power.

Problems of this kind are quite common in the theories we shall discuss in the next chapter. Clearly one could compute the large number 37^{13} and find its remainder when divided by 17. However, by congruences we proceed in much simpler fashion as follows. We observe first that

$$37 \equiv 3 \text{ (mod 17)}$$

By squaring this congruence, one finds

$$37^2 \equiv 9 \text{ (mod 17)}$$

and by repetition

$$37^4 \equiv 81 \equiv -4 \text{ (mod 17)}$$

Squaring again, one finds
$$37^8 \equiv 16 \equiv -1 \pmod{17}$$

By multiplying the congruences for the first, fourth, and eighth powers of 37 one obtains
$$37^{13} = 37 \cdot 37^4 \cdot 37^8 \equiv 3(-4)(-1) = 12 \pmod{17}$$

so the remainder is 12.

2. Compute the remainder of the expression
$$A = 531x^2 y^{11}, \qquad x = 31, y = 2$$
for the modulus 7.

One finds for the modulus 7

$$x = 31 \equiv 3, \qquad y^3 = 8 \equiv 1$$
$$x^2 \equiv 9 \equiv 2, \qquad y^9 \equiv 1$$
$$531 \equiv -1, \qquad y^{11} \equiv 4 \cdot 1 \equiv 4$$

Consequently
$$A \equiv -1 \cdot 2 \cdot 4 = -8 \equiv 6 \pmod{7}$$

3. Let
$$f(x) = 3x^7 - 41x^2 - 91x$$
and find $f(11) \pmod{13}$.

One sees that
$$41 \equiv 2, \qquad 91 \equiv 0 \pmod{13}$$
so that for any x
$$f(x) \equiv 3x^7 - 2x^2 \pmod{13}$$

Furthermore in this case

$$x = 11 \equiv -2$$
$$x^2 \equiv 4$$
$$x^4 \equiv 16 \equiv 3 \qquad \pmod{13}$$
$$x^7 \equiv -2 \cdot 4 \cdot 3 \equiv 2$$

so that
$$f(11) \equiv 3 \cdot 2 - 2 \cdot 4 = -2 \equiv 11 \pmod{13}$$

So far we have indicated only those rules for congruences corresponding to those that are familiar for equations. We shall now supplement this by deriving a number of properties for congruences that do not have an analogue among the properties of equations.

Almost trivial is

THEOREM 9-4. If

$$a \equiv b \pmod{m}$$

then one also has

$$a \equiv b \pmod{d}$$

where d is any divisor of the modulus m.

Clearly, if $a - b$ is divisible by m, it is divisible by any divisor d of m. For instance, one has

$$23 \equiv -1 \pmod{12}$$

and therefore

$$23 \equiv -1 \pmod{4}, \qquad 23 \equiv -1 \pmod{3}$$

Another fact that is often used in computations with congruences is the following:

THEOREM 9-5. When a congruence holds for two different moduls, it holds for their least common multiple. If

$$a \equiv b \pmod{m_1}, \qquad a \equiv b \pmod{m_2}$$

then

$$a \equiv b \pmod{M}, \qquad M = [m_1, m_2]$$

Conversely, the last congruence implies each of the first two.

The proof is an immediate consequence of the fact that when the difference $a - b$ is divisible both by m_1 and m_2, it is divisible by their l.c.m. M. Clearly the rule extends to an arbitrary number of moduls. In the example

$$37 \equiv 109 \pmod{8}, \qquad 37 \equiv 109 \pmod{12}$$

it follows that

$$37 \equiv 109 \pmod{24}$$

The converse is a consequence of theorem 9-4.

Let us state separately a special application of theorem 9-5 that appears commonly:

THEOREM 9-6. When a set of congruences

$$a \equiv b \pmod{m_i} \qquad i = 1, 2, \cdots, k \tag{9-8}$$

holds where the moduls m_i are relatively prime in pairs, then one also has the same congruence for the product of the moduls

$$a \equiv b \pmod{m_1 m_2 \cdots m_k} \tag{9-9}$$

and conversely from (9-9) each of the congruences (9-8) follows.

We need only to recall that the l.c.m. of relatively prime numbers is equal to their product.

Theorem 9-6 is often useful in reducing the study of congruences to the case of moduls that are powers of primes. If the modul has the prime factorization

$$m = p_1^{\alpha_1} p_2^{\alpha_2} \cdots p_k^{\alpha_k}$$

then the congruence

$$a \equiv b \pmod{m} \tag{9-10}$$

implies each one of the congruences

$$a \equiv b \pmod{p_i^{\alpha_i}} \tag{9-11}$$

and these, in turn, together imply (9-10). One has for instance

$$730 \equiv 10 \pmod{180}$$

and therefore also

$$730 \equiv 10 \pmod{2^2}, \pmod{3^2}, \pmod{5}$$

and conversely this system of congruences is equivalent to the original.

The final rules we wish to establish refer to the division of a congruence by a number. We have seen in theorem 9-2 that in a congruence both sides may be multiplied by the same integer. Now let us consider conversely when one can cancel a common factor on both sides. This is not always possible as the following example shows. In the congruence

$$36 \equiv 92 \pmod{8}$$

the numbers on both sides are divisible by 4, but if this factor is canceled, there remains

$$9 \equiv 23 \pmod{8}$$

which is incorrect

Let us see how the cancellation rule must be modified. When a congruence

$$ak \equiv bk \pmod{m}$$

holds, it means that the difference $ak - bk$ must be divisible by m so that

$$(a - b)k = lm \qquad (9\text{--}12)$$

where l is some integer. We assume that k and m have the g.c.d. $d = (k, m)$ and divide (9–12) by it to obtain

$$(a - b)\frac{k}{d} = l\frac{m}{d}$$

Here the two numbers k/d and m/d are relatively prime, and since the product on the left is divisible by m/d, one concludes that $a - b$ must be divisible by m/d, in other words

$$a \equiv b \left(\operatorname{mod} \frac{m}{d}\right)$$

We can therefore state:

THEOREM 9–7. In a congruence

$$ak \equiv bk \pmod{m}$$

the common factor k can be canceled

$$a \equiv b \left(\operatorname{mod} \frac{m}{d}\right)$$

provided the modulus is divided by the greatest common divisor d of k and m.

In the previous example

$$36 \equiv 92 \pmod{8}$$

cancellation by 4 gives, according to this rule,

$$9 \equiv 23 \pmod{2}$$

Similarly, in the congruence

$$220 \equiv 1{,}180 \pmod{96}$$

both sides have the common factor $k = 20$, and $d = (20, 96) = 4$. Consequently, after cancellation with 20, there remains

$$11 \equiv 59 \pmod{24}$$

Again, the theorem 9–7 has two special cases that are so important we mention them separately:

THEOREM 9–8.   In a congruence

$$ak \equiv bk \pmod{m}$$

the factor k may be canceled

$$a \equiv b \pmod{m}$$

provided k is relatively prime to the modul m.   For instance, in the congruence

$$27 \equiv 102 \pmod{25}$$

one can cancel by 3

$$9 \equiv 34 \pmod{25}$$

since 3 is relatively prime to the modul.

Finally:

THEOREM 9–9.   If in a congruence

$$a \equiv b \pmod{m}$$

the three numbers a, b, and m are divisible by a number d, then

$$\frac{a}{d} \equiv \frac{b}{d} \left(\text{mod} \, \frac{m}{d} \right)$$

Problems.

1. Add, subtract, multiply, and square the two congruences

$$31 \equiv -7, \quad 3 \equiv 22 \quad \pmod{19}$$

and check the results.

2. Compute the least positive residue of each of the numbers

(a) $2^{14} \pmod{17}$ (b) $11^{35} \pmod{13}$
(c) $2^{21} \pmod{11}$ (d) $3^{100} \pmod{5}$

3. Compute the residues of $f(2)$ and $f(13)$ (mod 12) when

$$f(x) = 73x^9 - 111x^7 + 32x - 14$$

4. Find the residue (mod 19) of the expression

$$B = 31x^2y + 17y^4x^5, \qquad x = 11, \ y = 24$$

5. Compute the remainders of the numbers

$$2! = 1 \cdot 2, \qquad 3! = 3 \cdot 2 \cdot 1, \qquad 4! = 4 \cdot 3 \cdot 2 \cdot 1, \ \ldots$$

and in general the remainder of $n!$ for the modulus $n + 1$ up to $n = 10$ and try to establish a general rule.

6. In the following congruences cancel the common factors on both sides:

(a) $264 \equiv 1{,}224$ (mod 48)
(b) $45 \equiv 150$ (mod 7)
(c) $168 \equiv -48$ (mod 72)

9–5. Casting out nines.

Until now we have mainly compiled rules for handling congruences, and the time has come to touch upon some simple applications to illustrate their usefulness. Towards the end of the first section of the *Disquisitiones* Gauss points out how one can, by means of congruences, derive general methods for checking numerical computations. Such checks are of ancient origin and may have been obtained from India by the Arabs together with the Hindu numerals. They occur in many of the Arab reckoning manuals, for instance, in the influential works of al-Khowarizmi and al-Karkhi, and so they came into general use in Europe in the Middle Ages.

These checks were particularly useful at a time when familiarity with arithmetic manipulations was not as widespread nor as thorough as at present. Furthermore, in computations on the abacus or casting on the lines, once the calculation was completed there remained no permanent record whose details could be rechecked. Nowadays these control methods, even the simplest and the most common one, casting out nines, have largely gone out of use and are no longer explained in the elementary texts in arithmetic. Occasionally we check our computations by the inverse operations, for instance, subtraction by adding the subtracted number to the difference, or division by multiplying back again, but in most cases the check is performed simply by going over each individual step in the calculation again.   However as anyone who has spent some time at numerical computations will realize, one is apt to succumb to the same pitfalls in repeating the procedure the second or third time, or in mechanical computation the machine may fail in the same manner as previously.   Consequently for any large-scale computation it is, if not absolutely necessary, at least very desirable to have some independent checking method for the results.

Since we perform our computations in the decadic system of numbers, we shall limit our considerations to such systems, but, as we will see, there is no difficulty in extending the results to arbitrary base numbers other than 10.   Let

$$N = (a_n, a_{n-1}, \cdots, a_1, a_0) \tag{9–13}$$
$$= a_n 10^n + a_{n-1} 10^{n-1} + \cdots + a_2 100 + a_1 10 + a_0$$

denote a number written in the decadic system so that the digits a_i may have values from 0 to 9.   It is simple to find the remainder of N when divided by divisors of the base number.   For instance, since 2 divides 10 and all powers of 10, it follows from (9–13) that

$$N \equiv a_0 \pmod 2$$

Consequently, N is divisible by 2 only when the last digit is divisible by 2, hence when a_0 has one of the values $a_0 = 0, 2, 4, 6, 8$. Similarly, since 4 divides 100 and all higher powers of 10, one has

$$N \equiv a_1 10 + a_0 \pmod 4$$

so that N is divisible by 4 only when the number represented by the two last digits is divisible by 4.   For example, the number

$$N = 7{,}342 \equiv 42 \equiv 2 \pmod 4$$

is not divisible by 4.   Equally simple and familiar are the rules for divisibility by 5 or 25.   One sees that

$$N \equiv a_0 \pmod 5$$

so that N is divisible by 5 only when $a_0 = 0$ or 5; one also has

$$N \equiv a_1 10 + a_0 \pmod{25}$$

so a number is divisible by 25 only when it ends in 00, 25, 50, or 75.

More interesting are the rules one can derive for the remainders and divisibility by other numbers that are relatively prime to 10. We begin by considering the number N in (9–13) for the modul $m = 9$. Since one has

$$10 \equiv 1 \pmod{9}$$

it follows that

$$10^2 \equiv 1, \qquad 10^3 \equiv 1, \quad \cdots \quad \pmod{9}$$

so that we find from (9–13)

$$N \equiv a_0 + a_1 + \cdots + a_n \pmod{9} \qquad (9\text{–}14)$$

This congruence expresses the basis for the process of casting out nines. It shows that by division with nine a number has the same remainder as the sum of the digits. One may notice that on an abacus or by computations on the lines this sum of the digits is a number that appears naturally since it is the number of counters or *jetons* that one uses to represent the number.

When the rule (9–14) is applied to find the remainder of a number with respect to the divisor 9, the sum of the digits may itself be a fairly large number, which one can reduce further by repeated application of the same rule. For instance,

$$N = 39{,}827{,}437 \equiv 3 + 9 + 8 + 2 + 7 + 4 + 3 + 7$$

$$= 43 \equiv 4 + 3 = 7 \pmod{9}$$

One concludes immediately from the congruence (9–14): A number is divisible by 9 only if the sum of its digits is divisible by 9.

Example.

The number

$$N = 234{,}648 \equiv 2 + 3 + 4 + 6 + 4 + 8 = 27 \equiv 0 \pmod{9}$$

is divisible by 9.

Since the congruence (9–14) also will hold for the divisor 3 of 9, exactly the same rules as for 9 apply for the remainders and divisibility by 3. For instance, a number is divisible by 3 only when the sum of the digits is divisible by 3.

Example.

$$N = 874,326 \equiv 8 + 7 + 4 + 3 + 2 + 6 = 30 \equiv 3 \pmod 9$$

is a number divisible by 3 but not by 9.

Another number for which simple divisibility rules can be established is $m = 11$. In this case one verifies that

$$10 \equiv -1, \qquad 10^2 \equiv 1, \qquad 10^3 \equiv -1,$$
$$10^4 \equiv 1, \cdots \qquad (\mathrm{mod}\ 11)$$

and one concludes from (9–13)

$$N \equiv a_0 - a_1 + a_2 - a_3 + \cdots \pmod{11} \qquad (9\text{–}15)$$

Example.

$$N = 39,827,437 \equiv 7 - 3 + 4 - 7 + 2 - 8 + 9 - 3 = 1 \pmod{11}$$

so that this number is not divisible by 11.

Rules for remainders and divisibility by other numbers have been derived but they are less simple than those we have already obtained. Leonardo Fibonnaci in his *Liber abaci*, in addition to the rules for 9 and 11, also gives a rule for the number 7. As we have seen, these rules depend essentially on the behavior of the powers of 10 for the chosen modulus. When $m = 7$, one obtains successively

$$10 \equiv 3, \qquad 10^2 \equiv 2, \qquad 10^3 \equiv -1, \qquad 10^4 \equiv -3,$$
$$10^5 \equiv -2, \qquad 10^6 \equiv 1 \qquad (\mathrm{mod}\ 7)$$

Consequently one has

$$N \equiv a_0 + 3a_1 + 2a_2 - a_3 - 3a_4$$
$$- 2a_5 + a_6 \cdots \pmod 7 \qquad (9\text{–}16)$$

Example.

$N = 39{,}827{,}437 \equiv +7 + 3 \cdot 3 + 2 \cdot 4 - 7$
$$- 3 \cdot 2 - 2 \cdot 8 + 9 + 3 \cdot 3 = 13 \equiv 6 \ (\mathrm{mod}\ 7)$$

shows that this number is not divisible by 7.

We shall now turn to the application of these residue rules to give checks for the correctness of arithmetic operations. These methods are based on the idea that when an operation of addition, subtraction, or multiplication has been performed on certain integers, the result must be correct also when considered as a congruence for an arbitrary modulus. For instance, let

$$c = ab \qquad (9\text{-}17)$$

be a product obtained by the multiplication of two numbers a and b. Then the congruence

$$c \equiv ab \ (\mathrm{mod}\ m) \qquad (9\text{-}18)$$

must hold for any modulus m. By selecting m as a number for which the residues may easily be computed by means of the preceding rules, the congruence (9-18) may be verified without much effort. If it should fail to hold, the multiplication (9-17) is not correct. On the other hand, if the congruence is fulfilled, the result (9-17) is not necessarily correct, but the chance of an error is considerably reduced.

When casting out nines, one uses the modulus $m = 9$.

Examples.

1. Let us take the multiplication (9-17) when

$$a = 8{,}297, \qquad b = 3{,}583, \qquad c = 29{,}728{,}151 \qquad (9\text{-}19)$$

Here one finds

$$a \equiv 8 + 2 + 9 + 7 = 26 \equiv -1$$
$$b \equiv 3 + 5 + 8 + 3 = 19 \equiv 1 \qquad\qquad (\mathrm{mod}\ 9)$$
$$c \equiv 2 + 9 + 7 + 2 + 8 + 1 + 5 + 1 = 35 \equiv -1$$

Consequently

$$ab \equiv -1, \qquad c \equiv -1 \qquad (\mathrm{mod}\ 9)$$

as one should expect. We shall also check the multiplication (9–19) for the modulus 11. Then by (9–15)

$$a \equiv 7 - 9 + 2 - 8 \equiv 3$$

$$b \equiv 3 - 8 + 5 - 3 \equiv -3 \qquad \text{(mod 11)}$$

$$c \equiv 1 - 5 + 1 - 8 + 2 - 7 + 9 - 2 \equiv 2$$

and therefore

$$ab \equiv 3(-3) \equiv 2, \qquad c \equiv 2 \qquad \text{(mod 11)}$$

2. Let us assume that in (9–17) one has

$$a = 7{,}342, \qquad b = 2{,}591, \qquad c = 19{,}032{,}122 \qquad (9\text{–}20)$$

By casting out nines, one finds

$$a \equiv 7, \qquad b \equiv 8, \qquad c \equiv 2 \qquad \text{(mod 9)}$$

which checks, since

$$ab \equiv 2 \ (\text{mod } 9)$$

But when one uses the modul 11, one finds from (9–20)

$$a \equiv 5, \qquad b \equiv 6, \qquad c \equiv 10 \qquad \text{(mod 11)}$$

and this indicates that there must be an error in the multiplication since

$$ab \equiv 30 \equiv 8 \not\equiv c \ (\text{mod } 11)$$

By performing the multiplication a second time one finds that the correct value should have been

$$c^* = 19{,}023{,}122$$

This illustrates the fact that casting out nines will not catch the rather common error of two digits having been interchanged.

These checking methods may be used analogously for addition and subtraction but they are of lesser importance since these operations may be so easily repeated. On the other hand, for division the checks are quite convenient. When an integer a is divided by b with the incomplete quotient q and the remainder r, the relation

$$a = qb + r$$

must hold for every modulus.

Example.

Let

$$a = 76,638,123, \qquad b = 37,547$$

By performing the division one finds

$$q = 2,041, \qquad r = 4,696$$

Casting out nines gives

$$a \equiv 0, \qquad b \equiv -1, \qquad q \equiv -2, \qquad r \equiv -2 \qquad (\text{mod } 9)$$

and this is correct since

$$qb + r \equiv (-2)(-1) - 2 \equiv 0 \ (\text{mod } 9)$$

similarly

$$a \equiv 1, \qquad b \equiv 4, \qquad q \equiv -5, \qquad r \equiv -1 \qquad (\text{mod } 11)$$

which again checks, since

$$qb + r \equiv 4(-5) - 1 \equiv 1 \ (\text{mod } 11)$$

With large figures it is more efficient to take larger moduls for check purposes. One may, for instance, use $m = 99$ since in this case

$$10^2 \equiv 1, \qquad 10^4 \equiv 1, \ \cdots \qquad (\text{mod } 99)$$

and from (9–13) one obtains

$$N \equiv a_0 + 10a_1 + a_2 + 10a_3 + \cdots (\text{mod } 99)$$

This means that one finds the remainder (mod 99) by splitting N up into two digit numbers and taking their sum. For instance, when

$$N = 7,342,948$$

one has

$$N \equiv 48 + 29 + 34 + 7 \equiv 19 \ (\text{mod } 99)$$

Similarly one finds for the modul $m = 101$

$$N \equiv a_0 + 10a_1 - (a_2 + 10a_3) + (a_4 + 10a_5) - \cdots (\text{mod } 101)$$

hence in the example we just used

$$N \equiv 48 - 29 + 34 - 7 = 46 \ (\text{mod } 101)$$

Let us check the multiplication

$$728{,}223 \times 5{,}535{,}064 = 4{,}030{,}760{,}911{,}272$$

by these two moduls. For the three numbers one finds in the given order (mod 99)

23	64	72
+82	+50	+12
+72	+53	+91
$\overline{177} \equiv 78$	$+\ 5$	+60
$\equiv -21$	$\overline{172} \equiv 73$	+07
	$\equiv -26$	+03
		$+\ 4$
		$\overline{249} \equiv 51$

The fact that

$$(-21)(-26) \equiv 51 \pmod{99}$$

points to a correct result. Had one used the modulus 101, the check would have looked as follows:

23	64	72
−82	−50	−12
+72	+53	+91
$\overline{13}$	$-\ 5$	−60
	$\overline{62}$	$+\ 7$
		$-\ 3$
		$+\ 4$
		$\overline{99} \equiv -2$

And here one has

$$13 \cdot 62 \equiv -2 \pmod{101}$$

In number theory one often runs into computations with very large numbers, exceeding even the capacity of the machines, so that multiplications and divisions may have to be performed in installments. Furthermore, there may be chains of computations where the result of one step enters into the next. This, for instance, is the case in some of the methods for deciding whether a number is a prime or not. Under such circumstances it is particularly important to have efficient checks, and one introduces moduls for even higher numbers than those just mentioned. It is convenient to take $m = 999$, $m = 9,999, \ldots$ since this leads to a simple addition of the digits of the numbers in groups of three, four, and so on, or one may also take $m = 1,001, \ldots$ and add and subtract such groups of digits alternatingly. By means of adding machines these checks may be performed with relatively small effort in comparison with the work involved in a complete repetition of the operation.

Problems.

1. In Arab and medieval European arithmetics one finds checks for $m = 7$, 9, 11 and also for $m = 13$ and $m = 19$. Determine the form of the residue rules for the two last moduls.

2. Why does $m = 17$ not give a simple rule?

3. Try to give a criterion for the divisibility of a number by 37.

4. Check the following multiplications and divisions by several of the rules established above:

(a) $14,745 \times 19,742 = 291,095,790$

(b) $52,447 \times 81,484 = 4,279,531,348$

(c) $24,726,928,309 = 3,569,644 \times 6,927 + 4,321$

(d) $41,587^2 = 1,729,478,569$

5. If a is a number in the decadic system and b the number with the same digits in the reverse order, prove that $a - b$ is divisible by 9.

6. What rule corresponds to the simultaneous use of the moduls 7, 11, and 13?

Bibliography

Gauss, C. F.: *Disquisitiones arithmeticae*, Leipzig, 1801. French translation by A. C. M. Poullet-Delisle: *Recherches arithmetiques*, Paris, 1807.

CHAPTER 10

ANALYSIS OF CONGRUENCES

10–1. Algebraic congruences. The congruences, as we have already indicated several times, have many properties in common with equations and this analogy we shall now pursue further. In equation theory one tackles the problem of finding the *roots* of an *algebraic equation*

$$f(x) = 0$$

i.e., the numbers x that satisfy this condition, where $f(x)$ is some given polynomial. Similarly, in the theory of congruences one can propose the problem of finding those integers x that fulfill a certain congruence

$$f(x) \equiv 0 \pmod{m} \tag{10–1}$$

for some modul. Since we deal only with integers, we must in this case suppose that $f(x)$ is a polynomial with integral coefficients.

As an example, let us take the congruence

$$f(x) = x^3 + 5x - 4 \equiv 0 \pmod{7} \tag{10–2}$$

It is satisfied when $x = 2$ since

$$f(2) = 14 \equiv 0 \pmod{7}$$

As for equations, we say that $x = 2$ is a *root* or *solution* of the congruence. But since congruent values of x will give congruent values of the polynomial, as we mentioned earlier, any value of x congruent to 2 (mod 7) must also be a solution. In the theory of congruences it is therefore agreed to consider all values

$$x \equiv 2 \pmod{7}$$

234

as *one* solution. In the general case (10–1) the situation is the same. When a solution $x = x_0$ has been found, all values x for which

$$x \equiv x_0 \pmod{m}$$

are also solutions and by convention we consider them as a *single* solution.

As a consequence, to find all solutions of a congruence (10–1) we need only try all values $0, 1, \ldots, m - 1$ [or the numbers in any other complete residue system (mod m)] and determine which of them satisfy the congruence; this gives us the total number of different solutions. In the example (10–2) we should try the numbers from 0 to 6 or, more conveniently, the numbers from -3 to $+3$. One finds that there is only the single solution $x \equiv 2$.

The number of solutions of a congruence may vary considerably; there may be none or the number may even greatly exceed the degree of the congruence.

Examples.

1. The congruence

$$x^2 + 5 \equiv 0 \pmod{11}$$

has no solutions as one establishes by trying the eleven values $0, \pm1, \pm2, \pm3, \pm4, \pm5$.

2. The congruence

$$x^3 - 2x + 6 \equiv 0 \pmod 5$$

has the two solutions

$$x \equiv 1, \qquad x \equiv 2 \qquad \pmod 5$$

3. The congruence

$$x^3 \equiv 0 \pmod{27}$$

has nine solutions

$$x \equiv 0, \qquad x \equiv \pm3, \qquad x \equiv \pm6, \qquad x \equiv \pm9, \qquad x \equiv \pm12 \qquad \pmod{27}$$

The theory of algebraic congruences is an interesting but quite complicated and difficult field, in which many investigations have been made during the last century. A few essential facts about special congruences will be derived subsequently since they enter into some of the applications of the theory of congruences that we wish to make. For the moment we shall be content to have introduced the basic concepts.

Problems.

Find all solutions of the following algebraic congruences:

1. $x^2 \equiv 5 \pmod{11}$

2. $x^2 \equiv 4 \pmod{15}$

3. $x^2 \equiv 1 \pmod{32}$

4. $x^3 \equiv 0 \pmod{25}$

5. $x^3 - 3x^2 - 3 \equiv 0 \pmod{13}$

6. $x^4 - 2x + 5 \equiv 0 \pmod{7}$

7. $x^3 - 3x^2 + 7x + 2 \equiv 0 \pmod{12}$

8. $x^{10} \equiv 1 \pmod{11}$

10–2. Linear congruences. The simplest congruences are those of *first degree,* or *linear congruences.*

$$ax \equiv b \pmod{m} \tag{10-3}$$

Before we proceed to the general method for solving such congruences we shall give a few examples to illustrate that there are various possibilities which may occur.

Examples.

1. The congruence

$$7x \equiv 3 \pmod{12}$$

has the single solution

$$x \equiv 9 \pmod{12}$$

as one concludes by trying out the integers from 0 to 11.

2. The congruence

$$12x \equiv 2 \pmod{8}$$

is found to have no solution.

3. Finally the congruence

$$6x \equiv 9 \pmod{15}$$

has three solutions

$$x \equiv 4, \qquad x \equiv 9, \qquad x \equiv 14 \qquad \pmod{15}$$

We now return to the general linear congruence (10–3). According to the definition of congruences, this equation means that there shall exist some integer y such that

$$ax - b = my$$

or

$$ax - my = b \tag{10-4}$$

This shows that the solution of a congruence (10–3) is equivalent to the solution of a linear indeterminate equation (10–4), and since we have already analyzed such equations quite exhaustively, the results may be transferred directly to congruences.

We recall first that an indeterminate equation (10–4) has a solution only if the greatest common divisor of the coefficients of x and y also divides the constant term b. Therefore we can state:

THEOREM 10–1. A linear congruence (10–3) is solvable only when the greatest common divisor

$$d = (a, m)$$

divides b.

In the first example given above

$$d = (7, 12) = 1$$

divides $b = 3$ so that the congruence is solvable. Similarly in the third example

$$d = (6, 15) = 3$$

divides $b = 9$ so that there are solutions. But in the second example

$$d = (12, 8) = 4$$

does not divide $b = 2$ so that no solution can exist, as we found directly.

Let us consider the general indeterminate equation (10–4) and suppose that d divides b so that it is solvable. We can cancel d in each term and obtain

$$\frac{a}{d}x - \frac{m}{d}y = \frac{b}{d} \tag{10–5}$$

This equation, as one sees, corresponds to the linear congruence

$$\frac{a}{d}x \equiv \frac{b}{d}\left(\bmod \frac{m}{d}\right) \tag{10–6}$$

In (10–5) the coefficients of x and y are now relatively prime, and we can solve the equation by means of our previous methods. We recall that if x_0 and y_0 is an arbitrary solution of (10–5), the general solution has the form

$$x = x_0 + \frac{m}{d} \cdot t, \qquad y = y_0 + \frac{m}{d} \cdot t$$

where t is an arbitrary integer. This gives us as the general solution of the congruences (10–6) and (10–3)

$$x \equiv x_0 \left(\operatorname{mod} \frac{m}{d} \right) \qquad (10\text{–}7)$$

There is one further remark that must be made. In connection with congruences we agreed that the different solutions were the numbers satisfying the congruence and not congruent to each other (mod m). The numbers (10–7) are not all congruent (mod m). If we select x_0, as we may, to be a positive integer less than m/d, all the numbers

$$x_0, \quad x_0 + \frac{m}{d}, \quad x_0 + 2\frac{m}{d}, \quad \cdots, \quad x_0 + (d-1)\frac{m}{d} \qquad (10\text{–}8)$$

satisfy the congruence and are incongruent (mod m), since they are less than m. Then the d numbers (10–8) define different solutions of the original congruence (10–3). To summarize:

THEOREM 10–2. A congruence

$$ax \equiv b \pmod{m}$$

is solvable only if the greatest common divisor $d = (a, m)$ divides b, and when this is the case there are d solutions given by (10–8). When a and m are relatively prime, the congruence has a single solution.

In the first example above we had $d = 1$ so that there was one solution. In the third example we had $d = 3$, and when this factor was canceled the congruence became

$$2x \equiv 3 \pmod 5$$

with the general solution

$$x \equiv 4 \pmod 5$$

Corresponding to (10–8) it follows that since the modul in the original congruence was 15, one has altogether three solutions

$$x \equiv 4, \qquad x \equiv 9, \qquad x \equiv 14 \qquad (\text{mod } 15)$$

Examples.

1. The congruence
$$36x \equiv 8 \ (\text{mod } 102)$$
has no solution since
$$d = (36, 102) = 6$$
does not divide $b = 8$.

2. The congruence
$$19x \equiv 1 \ (\text{mod } 140)$$
has a single solution since $d = 1$. To obtain it we solve the equation
$$19x - 140y = 1$$
by means of our previous procedure based upon Euclid's algorism

$$
\begin{array}{c|c}
140 = 19 \cdot 7 + 7 & 59 \\
19 = \ 7 \cdot 2 + 5 & 8 \\
7 = \ 5 \cdot 1 + 2 & 3 \\
5 = \ 2 \cdot 2 + 1 & 2 \\
& 1
\end{array}
$$

The solution is therefore
$$x \equiv 59 \ (\text{mod } 140)$$

3. Finally in the example
$$144x \equiv 216 \ (\text{mod } 360)$$
one has $d = 72$, and this number divides $b = 216$ so that there are 72 different solutions (mod 360). When the factor d is canceled in the congruence, there remains
$$2x \equiv 3 \ (\text{mod } 5)$$
which has the solution
$$x \equiv 4 \ (\text{mod } 5)$$

The 72 solutions of the original congruence are as in (10–8)

$$x \equiv 4, \qquad x \equiv 9, \qquad x \equiv 14, \ \cdots, \ \ x \equiv 359 \qquad (\text{mod } 360)$$

Problems.

Solve the congruences

1. $7x \equiv 3 \pmod{10}$ 4. $20x \equiv 7 \pmod{15}$

2. $15x \equiv 9 \pmod{12}$ 5. $315x \equiv 11 \pmod{501}$

3. $221x \equiv 111 \pmod{360}$ 6. $360x \equiv 3{,}072 \pmod{96}$

10–3. Simultaneous congruences and the Chinese remainder theorem. It is often required to find a number that has prescribed residues for two or several moduls. As an example, let us suppose that we wish to determine an integer x such that

$$x \equiv 5 \pmod{11}, \qquad x \equiv 3 \pmod{23} \qquad (10\text{–}9)$$

The first condition (10–9) states that

$$x = 5 + 11t$$

where t is some integer. In order that x shall satisfy the second congruence in (10–9), one must have

$$5 + 11t \equiv 3 \pmod{23}$$

or

$$11t \equiv -2 \pmod{23}$$

The solution of this congruence is obtained most simply by multiplying both sides by 2 so that

$$22t \equiv -t \equiv -4 \pmod{23}$$

or

$$t \equiv 4 \pmod{23}$$

The general form for t is therefore

$$t = 4 + 23u$$

where u is some integer. When this is substituted into the expression for x, one obtains

$$x = 5 + 11(4 + 23u) = 49 + 11 \cdot 23 \cdot u$$

so that the general solution of the two congruences (10–9) is

$$x \equiv 49 \pmod{11 \cdot 23}$$

The method used in this example is applicable in the general case of two congruences

$$x \equiv a \pmod{m}, \qquad x \equiv b \pmod{n} \qquad (10\text{--}10)$$

From the first of these, follows

$$x = a + mt$$

and the second shows that t must satisfy the condition

$$a + mt \equiv b \pmod{n}$$

or

$$mt \equiv b - a \pmod{n} \qquad (10\text{--}11)$$

According to the general rules we just derived, this linear congruence in t can only have a solution when the greatest common divisor $d = (m, n)$ divides $b - a$; in other words, the condition

$$a \equiv b \pmod{d}$$

must be fulfilled. When this is the case the congruence (10–11) may be divided by d

$$\frac{m}{d} t \equiv \frac{b - a}{d} \left(\bmod \frac{n}{d} \right) \qquad (10\text{--}12)$$

Let t_0 be some particular solution of this congruence and

$$x_0 = a + mt_0$$

the resulting special solution of (10–10). The general solution of (10–12) is then

$$t \equiv t_0 \left(\bmod \frac{n}{d} \right)$$

so that we can write

$$t = t_0 + u \frac{n}{d}$$

where u is some integer. The resulting general solution of the original congruence (10–10) is

$$x \equiv a + m\left(t_0 + \frac{n}{d}u\right) = x_0 + u\,\frac{mn}{d}$$

or

$$x \equiv x_0 \pmod{[m,\ n]}$$

since

$$[m,\ n] = \frac{mn}{d}$$

is the least common multiple of m and n.

When one considers a set of algebraic congruences for several moduls and x_0 is a number satisfying all of them, it is clear that if one adds any multiple of the l.c.m. of all moduls to x_0, the resulting number will also be a solution. Therefore, with several moduls it is agreed that the number of different solutions is given by the incongruent solutions for the l.c.m. of the moduls.

We summarize as follows:

THEOREM 10–3.   Two simultaneous congruences

$$x \equiv a \pmod{m}, \qquad x \equiv b \pmod{n}$$

are solvable only when

$$a \equiv b \pmod{(m,\ n)}$$

and then there is a single solution

$$x \equiv x_0 \pmod{[m,\ n]}$$

which may be found by the method given above.

Examples.

1. When

$$x \equiv 7 \pmod{42}, \qquad x \equiv 15 \pmod{51}$$

there is no solution since

$$d = (42,\ 51) = 3$$

and

$$7 \not\equiv 15 \pmod{3}$$

2. In the example

$$x \equiv 3 \pmod{14}, \qquad x \equiv 7 \pmod{16}$$

one has $d = 2$, and the condition for solvability is fulfilled; the answer is

$$x \equiv 87 \pmod{112}$$

When several simultaneous congruences are given

$$x \equiv a_1 \pmod{m_1}, \qquad x \equiv a_2 \pmod{m_2}, \qquad x \equiv a_3 \pmod{m_3}$$
$$(10\text{--}13)$$

the solution may be found by repeated applications of the method given above. One combines the first congruences and finds a single congruence

$$x \equiv x_0 \pmod{[m_1, m_2]}$$

which can replace them in (10–13). This in turn is solved in conjunction with the third, and so on. One sees that if there exists a solution of the congruences (10–13), there is only a single one, with respect to a modul that is the l.c.m. of all moduls m_i.

Example.

The following example is taken from the *Disquisitiones:*

$$x \equiv 17 \pmod{504}, \qquad x \equiv -4 \pmod{35}, \qquad x \equiv 33 \pmod{16}$$

When the two first congruences are solved, one finds

$$x \equiv 521 \pmod{2{,}520}$$

and when this is combined with the third, the result is

$$x \equiv 3{,}041 \pmod{5{,}040}$$

Many puzzle questions belong mathematically to the type of problems solved by simultaneous congruences. In Chap. 6 we mentioned the ancient problem of the woman with a basket of eggs. When the eggs were taken out two at a time, there was one left; similarly, when they were taken out 3, 4, 5, and 6 at a time, there was always one egg left, while at seven at a time, the count came out even. In mathematical terms this means that

$$x \equiv 1 \pmod{2, 3, 4, 5, 6}$$
$$x \equiv 0 \pmod{7}$$

where x is the number of eggs. Since, according to the first congruences, the number $x - 1$ is divisible by all numbers 2, 3, 4, 5, and 6, it is divisible by their l.c.m., which is 60, and the conditions become

$$x \equiv 1 \pmod{60}, \qquad x \equiv 0 \pmod{7}$$

When these simultaneous congruences are solved, one finds

$$x \equiv 301 \pmod{420}$$

The smallest number of eggs the basket could have contained is therefore $x = 301$.

The simple condition for the solvability of two simultaneous congruences given in theorem 10–3 can be extended to an arbitrary number of congruences as follows:

THEOREM 10–4. The necessary and sufficient condition for a set of simultaneous congruences

$$x \equiv a_i \pmod{m_i} \quad i = 1, 2, \cdots, r \quad (10\text{–}14)$$

to have a solution is that for any pair

$$a_i \equiv a_j \pmod{(m_i, m_j)} \tag{10–15}$$

and in this case, there is a single solution for the modulus

$$M_r = [m_1, \cdots, m_r]$$

which is the l.c.m. of the given ones.

We observe first that if the congruences (10–14) are to have a solution, any pair of them must be solvable so that according to theorem 10–3 it is necessary that the conditions (10–15) be fulfilled. To prove that these conditions are sufficient for the existence of a solution, we shall use the induction procedure. Theorem 10–3 states that the result is true for two congruences. We suppose, therefore, that the result is true when there are $r - 1$ congruences and from this we deduce it for r congruences. According to this assumption, there exists a solution

$$x_0 \equiv a_i \pmod{m_i} \quad i = 1, 2, \cdots, r - 1 \tag{10–16}$$

of the $r - 1$ first congruences in (10–14), and any other such solution x must be of the form

$$x \equiv x_0 \pmod{M_{r-1}}, \qquad M_{r-1} = [m_1, \cdots, m_{r-1}] \quad (10\text{–}17)$$

To have a solution of all congruences (10–14), one must at the same time satisfy

$$x \equiv a_r \pmod{m_r} \qquad (10\text{–}18)$$

From theorem 10–3 we conclude again that the congruences (10–17) and (10–18) can have a common solution only when

$$x_0 \equiv a_r \pmod{(M_{r-1}, m_r)} \qquad (10\text{–}19)$$

and in this case there is a single solution for the modulus

$$[M_{r-1}, m_r] = M_r$$

It remains, therefore, to show that the condition (10–19) is fulfilled for the x_0 we have found. We observe that according to theorem 5–9

$$(M_{r-1}, m_r) = ([m_1, \cdots, m_{r-1}], m_r) = [(m_1, m_r), \cdots, (m_{r-1}, m_r)]$$

so that the congruence (10–19) is equivalent to the set of congruences (theorem 9–5)

$$x_0 \equiv a_r \pmod{(m_i, m_r)} \qquad i = 1, 2, \ldots, r - 1$$

But these congruences are true, since one finds from (10–16) and (10–15)

$$x_0 \equiv a_i \pmod{(m_i, m_r)}, \qquad a_i \equiv a_r \pmod{(m_i, m_r)}$$

The special case where all moduls in the simultaneous congruences (10–14) are relatively prime in pairs occurs in many applications. According to theorem 10–4, there is a unique solution to these congruences for a modul that is equal to the product of all the given ones. Gauss introduces a special procedure already used previously by Euler for the determination of the solution. The method, however, is ancient and occurs in the works of severa early mathematicians. The first known source is the *Arithmetic* of the Chinese writer Sun-Tse, probably around the beginning of our era, and the resulting formula is often termed the *Chinese remainder theorem*.

We begin by forming the product

$$M = m_1 m_2 \cdots m_r$$

of the relatively prime moduls in the set of congruences (10–14). When M is divided by m_1, the quotient

$$\frac{M}{m_1} = m_2 \cdots m_r$$

is a number divisible by all other moduls and relatively prime to m_1. Similarly the number

$$\frac{M}{m_i} = m_1 \cdots m_{i-1} m_{i+1} \cdots m_r$$

is divisible by all moduls except m_i, to which it is relatively prime. For each i one can, consequently, solve the linear congruence

$$b_i \frac{M}{m_i} \equiv 1 \pmod{m_i} \tag{10–20}$$

The Chinese remainder theorem may then be stated:

THEOREM 10–5. Let a set of simultaneous congruences (10–14) be given for which the moduls m_i are relatively prime. For each i one determines b_i through the linear congruence (10–20). The solution of the set of congruences is then

$$x \equiv a_1 b_1 \frac{M}{m_1} + a_2 b_2 \frac{M}{m_2} + \cdots + a_r b_r \frac{M}{m_r} \pmod{M} \tag{10–21}$$

The verification of the solution (10–21) is immediate. For instance, to see that it satisfies the first congruence (10–14) we recall that m_1 divides all M/m_i except M/m_1 so that

$$x \equiv a_1 b_1 \frac{M}{m_1} \equiv a_1 \pmod{m_1}$$

The other congruences (10–14) follow by similar arguments.

The example used by Sun-Tse corresponds to the three congruences

$$x \equiv 2 \ (\text{mod } 3), \qquad x \equiv 3 \ (\text{mod } 5), \qquad x \equiv 2 \ (\text{mod } 7)$$

Here $M = 105$ and

$$\frac{M}{m_1} = 35, \qquad \frac{M}{m_2} = 21, \qquad \frac{M}{m_3} = 15$$

The set of linear congruences

$$35b_1 \equiv 1 \ (\text{mod } 3), \qquad 21b_2 \equiv 1 \ (\text{mod } 5), \qquad 15b_3 \equiv 1 \ (\text{mod } 7)$$

has the solutions

$$b_1 = 2, \qquad b_2 = 1, \qquad b_3 = 1$$

so that one finds, according to (10–21),

$$x \equiv 2 \cdot 2 \cdot 35 + 3 \cdot 1 \cdot 21 + 2 \cdot 1 \cdot 15 \equiv 233 \ (\text{mod } 105)$$

or

$$x \equiv 23 \ (\text{mod } 105)$$

Congruences represent a very convenient tool in many calendar questions, such as determination of Easter dates, the day of the week of a particular date, and so on. Gauss illustrates the Chinese remainder theorem on a problem to find the years that have a certain period number with respect to the solar and lunar cycle and the Roman indiction. Similar problems with respect to the planetary cycles occur earlier by Brahmagupta.

In the formula (10–21) for the solution of congruences for relatively prime moduls, the multipliers

$$b_i \frac{M}{m_i}$$

depend only on the numbers m_i. If, therefore, one has to solve several sets of congruences, all with the same moduls, the expression (10–21) is particularly convenient since one need not recalculate the multipliers for each set.

As an example, let us take a play problem Leonardo discusses in the *Liber abaci*. A person is requested to think of some number.

Then he is asked what the remainders of the number are when it is divided by 5, 7, and 9, and on the basis of this information the number is divined.

Let us denote the unknown number by x and the three remainders by a_1, a_2, and a_3 so that

$$x \equiv a_1 \ (\text{mod } 5), \qquad x \equiv a_2 \ (\text{mod } 7), \qquad x \equiv a_3 \ (\text{mod } 9)$$

The moduls are relatively prime, and one has

$$M = 5 \cdot 7 \cdot 9 = 315$$

$$\frac{M}{m_1} = 63, \qquad \frac{M}{m_2} = 45, \qquad \frac{M}{m_3} = 35$$

The linear congruences

$$63b_1 \equiv 1 \ (\text{mod } 5), \qquad 45b_2 \equiv 1 \ (\text{mod } 7), \qquad 35b_3 \equiv 1 \ (\text{mod } 9)$$

have the solutions

$$b_1 = 2, \qquad b_2 = 5, \qquad b_3 = 8$$

so that the Chinese remainder formula (10–21) yields

$$x \equiv 126a_1 + 225a_2 + 280a_3 \ (\text{mod } 315)$$

From this expression one obtains x, according to the remainders a_1, a_2, a_3 indicated. Only when the number is required to be less than 315 is there a unique solution.

We conclude these investigations with a remark that will be applied later. Let us suppose that in solving a problem for a modul m_1 the number x to be determined has s_1 admissible values

$$x \equiv a_1, a_2, \cdots, a_{s_1} \ (\text{mod } m_1)$$

Similarly for the modul m_2 there are s_2 values

$$x \equiv b_1, b_2, \cdots, b_{s_2} \ (\text{mod } m_2)$$

When m_1 and m_2 are relatively prime, each value a_i (mod m_1) may be combined with an arbitrary value b_j (mod m_2) so that there exists a total of $s_1 s_2$ admissible values (mod $m_1 m_2$). In general, one sees that if there are r moduls m_1, m_2, ..., m_r, all relatively prime, and $s_1, s_2, \ldots, s_r$ possible values for x for each modul, there will be $s_1 s_2 \cdots s_r$ possible values for x for the product modul $m_1 m_2 \cdots m_r$.

Problems.

1. The basket of eggs problem is often given in the form: When the eggs are taken out 2, 3, 4, 5, 6 at a time, there remain respectively 1, 2, 3, 4, 5 eggs, while the number comes out even when they are taken out seven at a time. Find the smallest number of eggs there could have been in the basket. Brahmagupta discusses such a problem and makes the comment that it is a popular one.

2. Ancient Chinese problem. Three farmers divide equally the rice they have raised in common. They went to different markets where various basic weights were used, at one place 83 pounds, at another 110 pounds, and at the third 135 pounds. Each sold as much as he could in full measures, and when they came home one had 32 pounds left, another 70 pounds, and the third 30 pounds. How much rice had they raised together?

3. Ancient Chinese problem. Four labor gangs take over the construction of a dam, each contracting to take the same total number of workdays. The first gang consists of two men, the second has three, the third six, and the fourth twelve men. They complete their work as far as possible in full work by each gang, and then there remains one workday for one man for the first gang, two for the second, and five for the third and fourth. How many workdays did the whole project involve?

4. Regiomontanus. Find a number such that

$$x \equiv 3 \ (\text{mod } 10), \qquad x \equiv 11 \ (\text{mod } 13), \qquad x \equiv 15 \ (\text{mod } 17)$$

5. Euler. Find a number such that

$$x \equiv 3 \ (\text{mod } 11), \qquad x \equiv 5 \ (\text{mod } 19), \qquad x \equiv 10 \ (\text{mod } 29)$$

10–4. Further study of algebraic congruences. The methods we have just developed for simultaneous congruences may be applied to find the solutions of several algebraic congruences. If, for instance, one has two congruences

$$f(x) \equiv 0 \ (\text{mod } m), \qquad g(x) \equiv 0 \ (\text{mod } n)$$

and one wishes to find those x's which satisfy both at the same time, each congruence may be solved separately, and the two sets of roots for the moduls m and n may be combined as simultaneous congruences.

Example.

1. Let us take

$$x^3 - 2x + 3 \equiv 0 \pmod{7}$$
$$2x^2 \equiv 3 \pmod{15}$$

The first congruence is found to have a single solution

$$x \equiv 2 \pmod{7}$$

while the second has two solutions

$$x \equiv \pm 3 \pmod{15}$$

The simultaneous congruences

$$x \equiv 2 \pmod{7}, \qquad x \equiv 3 \pmod{15}$$

give

$$x \equiv 93 \pmod{105}$$

for the common solution of the two given congruences, and similarly another solution

$$x \equiv 72 \pmod{105}$$

is obtained from

$$x \equiv 2 \pmod{7}, \qquad x \equiv -3 \pmod{15}$$

2. We wish to find some x such that

$$12x \equiv 3 \pmod{15}, \qquad 10x \equiv 14 \pmod{8}$$

The first congruence, according to the theory of linear congruences, has the three solutions

$$x \equiv 4, \qquad x \equiv 9, \qquad x \equiv 14 \qquad \pmod{15}$$

while the second has two solutions

$$x \equiv 3, \qquad x \equiv 7 \qquad \pmod{8}$$

When these are combined, one obtains six solutions

$$\begin{aligned}
x &\equiv 19, & x &\equiv 79 \\
x &\equiv 39, & x &\equiv 99 \qquad \pmod{120} \\
x &\equiv 59, & x &\equiv 119
\end{aligned}$$

satisfying both congruences for the l.c.m. of the moduls.

By means of the results for simultaneous congruences, one can reduce the solution of an algebraic congruence to the case where the modul is a power of a prime. Let m be some number and

$$m = p_1^{\alpha_1} \cdot \cdots \cdot p_r^{\alpha_r}$$

its factorization into prime factors. An algebraic congruence

$$f(x) \equiv 0 \pmod{m} \qquad (10\text{--}22)$$

will hold only for those x's that at the same time satisfy each of the congruences

$$f(x) \equiv 0 \pmod{p_i^{\alpha_i}} \qquad i = 1, 2, \cdots, r \quad (10\text{--}23)$$

To solve the congruence (10–22) we can therefore determine the solutions of each congruence (10–23) separately and use the Chinese remainder theorem to obtain the values that satisfy them simultaneously.

Examples.

1. Let us take the congruence

$$x^3 - 7x^2 + 4 \equiv 0 \pmod{88}$$

Since 8 and 11 are the prime powers in the modul 88, we solve the congruence (mod 8) and (mod 11). In the first case one obtains three solutions

$$x \equiv 2, \qquad x \equiv 3, \qquad x \equiv 6 \qquad \pmod{8}$$

and in the second case there is a single solution

$$x \equiv 4 \pmod{11}$$

When these solutions are combined, one finds three solutions

$$x \equiv 26, \qquad x \equiv 59, \qquad x \equiv 70 \qquad \pmod{88}$$

of the original congruence.

2. Let us take

$$5x^2 + 1 \equiv 0 \pmod{189}$$

Here

$$189 = 3^3 \cdot 7$$

so that we solve the congruence (mod 3^3) and (mod 7). One finds, respectively,

$$x \equiv \pm 4 \pmod{27}, \qquad x \equiv \pm 2 \qquad \pmod{7}$$

and the combination of these gives four solutions of the given congruence

$$x \equiv \pm 23, \qquad x \equiv \pm 58 \qquad \pmod{189}$$

We have just seen how the solution of general algebraic congruences may be derived from congruences with a prime-power modulus.   One can go one step further and give a method to solve congruences for prime-power moduls by means of congruences for a prime modul.   We shall make the procedure clear on two examples.

Examples.

1. Let us first consider the congruence

$$f(x) = x^2 - 7x + 2 \equiv 0 \pmod 5 \tag{10–24}$$

By trial one finds the solutions

$$x \equiv 3 \quad \text{and} \quad x \equiv -1 \pmod 5 \tag{10–25}$$

In the second step we take the same congruence (10–24) for the modulus 25

$$f(x) = x^2 - 7x + 2 \equiv 0 \pmod{5^2} \tag{10–26}$$

It is clear that a solution of this congruence must be found among the numbers (10–25) so that we can put

$$x = 3 + 5t \quad \text{or} \quad x = -1 + 5u \tag{10–27}$$

To determine t and u we obtain from (10–26) by substitution of these values respectively

$$f(x) = -10 - 5t + 25t^2 \equiv 0 \pmod{25}$$
$$f(x) = 10 - 45u + 25u^2 \equiv 0$$

and this reduces to

$$t \equiv -2, \quad u \equiv -2 \pmod 5$$

Therefore, according to (10–27), we can write

$$t = -2 + 5s, \quad x = -7 + 25s \tag{10–28}$$
$$u = -2 + 5v, \quad x = -11 + 25v$$

so that the only solutions of the congruence (10–26) are

$$x \equiv -7, \quad x \equiv -11 \pmod{25}$$

In the third step we take the congruence (10–24) for a modul that is the third power of 5

$$f(x) = x^2 - 7x + 2 \equiv 0 \pmod{125} \tag{10–29}$$

The solutions x must be of the form (10–28), and when they are substituted, one obtains

$$f(x) = 100 - 525s + 625s^2 \equiv 0 \pmod{125}$$
$$f(x) = 200 - 725v + 625v^2 \equiv 0$$

as the conditions s and v must satisfy. This reduces to

$$s \equiv 4, \qquad v \equiv 2 \ (\text{mod } 5)$$

and it follows that

$$x \equiv 93, \qquad x \equiv 39 \ (\text{mod } 125)$$

are the only solutions of the congruence (10–29). The same process may be repeated indefinitely to obtain the solutions for moduls that are arbitrarily high powers of 5.

2. We begin by observing that the congruence

$$f(x) = x^3 - 8x^2 + 21x - 11 \equiv 0 \ (\text{mod } 7) \tag{10–30}$$

has the two solutions

$$x \equiv 2, \qquad x \equiv 3 \qquad (\text{mod } 7)$$

To find the solution of the congruence

$$f(x) = x^3 - 8x^2 + 21x - 11 \equiv 0 \ (\text{mod } 7^2) \tag{10–31}$$

we have to substitute

$$x = 2 + 7t, \qquad x = 3 + 7s \tag{10–32}$$

respectively. By the substitution of the second expression into (10–31), the condition reduces to the impossible congruence

$$7 \equiv 0 \ (\text{mod } 49)$$

so that we find no s that will give an x satisfying (10–31). When the first expression (10–32) is substituted, the congruence reduces to

$$t \equiv -1 \ (\text{mod } 7)$$

and correspondingly,

$$x \equiv -5 \ (\text{mod } 49)$$

is the only solution of (10–31). When this method is applied to the same congruence (10–30) (mod 7^3), one finds a single solution

$$x \equiv 93 \ (\text{mod } 343)$$

We shall conclude this study of algebraic congruences by establishing a few results that extend the analogies between equations and congruences. The first is:

THEOREM 10–6. When an algebraic congruence of degree n

$$f(x) \equiv 0 \ (\text{mod } m)$$

has a solution

$$x \equiv a_1 \ (\text{mod } m)$$

then one can write

$$f(x) \equiv (x - a_1)f_1(x) \ (\text{mod } m) \qquad (10\text{-}33)$$

where $f_1(x)$ is a polynomial of degree $n - 1$.

To prove the theorem we divide $f(x)$ by $x - a_1$ and find

$$f(x) = (x - a_1)f_1(x) + R$$

where R is some integer and the degree of $f_1(x)$ is one less than the degree of $f(x)$. By putting $x = a_1$ in this identity, we obtain

$$f(a_1) = R \equiv 0 \ (\text{mod } m)$$

so that the congruence (10-33) follows.

Examples.

1. Let us take the third-degree congruence

$$f(x) = x^3 - 7x^2 + 4 \equiv 0 \ (\text{mod } 88)$$

We have already found that this congruence has three solutions

$$x \equiv 26, \qquad x \equiv 59, \qquad x \equiv 70 \qquad (\text{mod } 88)$$

The division of $f(x)$ by $x - 26$ yields

$$x^3 - 7x^2 + 4 = (x - 26)(x^2 + 19x + 494) + 12{,}848$$

and since 12,848 is divisible by 88, one has

$$x^3 - 7x^2 + 4 \equiv (x - 26)(x^2 + 19x + 494) \ (\text{mod } 88)$$

The reader may determine the corresponding decompositions for the other roots of the congruence.

2. In the example of second degree

$$f(x) = 3x^2 + 7x - 2 \equiv 0 \ (\text{mod } 23)$$

one finds the root

$$x \equiv 3 \ (\text{mod } 23)$$

and the decomposition

$$3x^2 + 7x - 2 \equiv (x - 3)(3x + 16) \ (\text{mod } 23)$$

The remaining two theorems, it should be noted, hold only for congruences for a prime modul.

THEOREM 10-7. When the congruence

$$f(x) \equiv 0 \pmod{p} \qquad (10\text{-}34)$$

of nth degree for a prime modul p has r different roots

$$x \equiv a_1, \qquad x \equiv a_2, \cdots, x \equiv a_r \qquad (\text{mod } p)$$

one can write

$$f(x) \equiv (x - a_1)(x - a_2) \cdots (x - a_r)f_r(x) \pmod{p} \qquad (10\text{-}35)$$

where $f_r(x)$ is a polynomial of degree $n - r$.

The proof is based upon theorem 10-6, which shows first that one can write

$$f(x) \equiv (x - a_1)f_1(x) \pmod{p} \qquad (10\text{-}36)$$

where $f_1(x)$ is of degree $n - 1$. But since a_2 is also a root of (10-34), we must have

$$f(a_2) \equiv (a_2 - a_1)f_1(a_2) \equiv 0 \pmod{p}$$

Here we use the fact that when a prime divides a product it must divide one of the factors. The difference $a_2 - a_1$ is not divisible by p since a_1 and a_2 were different roots, so that we conclude

$$f_1(a_2) \equiv 0 \pmod{p}$$

According to theorem 10-6 we can write again

$$f_1(x) \equiv (x - a_2)f_2(x) \pmod{p}$$

where $f_2(x)$ is of degree $n - 2$; hence from (10-36)

$$f(x) \equiv (x - a_1)(x - a_2)f_2(x) \pmod{p}$$

For the third root a_3 of (10-34), one finds

$$f(a_3) \equiv (a_3 - a_1)\ a_3 - a_2)f_2(a_3) \equiv 0 \pmod{p}$$

and one concludes similarly that

$$f_2(a_3) \equiv 0 \pmod{p}$$

This gives

$$f_2(x) \equiv (x - a_3)f_3(x) \pmod{p}$$

where $f_3(x)$ is of degree $n - 3$, and

$$f(x) \equiv (x - a_1)(x - a_2)(x - a_3)f_3(x) \pmod{p}$$

The process may be continued until one arrives at the general decomposition (10–35).

Examples.

1. In the congruence of fourth degree

$$x^4 - 5x^3 - 5x - 1 \equiv 0 \pmod{7}$$

we have the roots

$$x \equiv 2, \qquad x \equiv 3 \qquad \pmod{7}$$

and corespondingly the decomposition

$$x^4 - 5x^3 - 5x - 1 \equiv (x - 2)(x - 3)(x^2 + 1) \pmod{7}$$

2. The congruence

$$x^4 - 1 \equiv 0 \pmod{5}$$

has the roots

$$x \equiv 1, \qquad x \equiv 2, \qquad x \equiv 3, \qquad x \equiv 4, \qquad \pmod{5}$$

and therefore

$$x^4 - 1 \equiv (x - 1)(x - 2)(x - 3)(x - 4) \pmod{5}$$

3. The congruence

$$3x^2 + 1 \equiv 0 \pmod{19}$$

has the solutions

$$x \equiv \pm5 \pmod{19}$$

and correspondingly one finds

$$3x^2 + 1 \equiv 3(x - 5)(x + 5) \pmod{19}$$

Our last result is due to the French mathematician Lagrange (1768), as Gauss observes in this connection in the *Disquisitiones.* Lagrange, of course, does not use the congruence terminology but the content of his theorem is as follows:

THEOREM 10–8.   A congruence

$$f(x) \equiv 0 \pmod{p}$$

for a prime modul p cannot have more different solutions than its degree, except in the trivial case where all coefficients in $f(x)$ are divisible by p.

Let us suppose that the degree of $f(x)$ is n and that

$$x \equiv a_1, \qquad x \equiv a_2, \ldots, x \equiv a_n \qquad (\text{mod } p)$$

are n different solutions of the congruence. From theorem 10–7 we conclude that

$$f(x) \equiv (x - a_1)(x - a_2) \cdots (x - a_n)F \quad (\text{mod } p)$$

where F is of zero degree, hence some integer. If there were some further solution

$$x \equiv a_{n+1} \ (\text{mod } p)$$

we would have

$$f(a_{n+1}) \equiv (a_{n+1} - a_1) \cdots (a_{n+1} - a_n)F \equiv 0 \ (\text{mod } p)$$

Here none of the differences $a_{n+1} - a_i$ are divisible by p since we deal with different solutions (mod p). The conclusion is that

$$F \equiv 0 \ (\text{mod } p)$$

and therefore identically

$$f(x) \equiv 0 \ (\text{mod } p)$$

which means that all coefficients in $f(x)$ are divisible by p.

Problems.

1. Find the common solutions to the congruences

    (a) $3x^2 - 7 \equiv 0 \ (\text{mod } 17)$

        $5x^2 - 2x - 3 \equiv 0 \ (\text{mod } 12)$

    (b) $3x \equiv 11 \ (\text{mod } 23)$

        $50x \equiv 2 \ (\text{mod } 32)$

    (c) $x^2 + 5 \equiv 0 \ (\text{mod } 27)$

        $3x + 1 \equiv 0 \ (\text{mod } 10)$

2. Solve the congruences

   (a) $x^3 - 3x - 8 \equiv 0 \pmod{60}$

   (b) $x^2 + 11 \equiv 0 \pmod{180}$

   (c) $x^2 + 2x + 7 \equiv 0 \pmod{75}$

3. Solve the following congruences and find the corresponding congruence factorizations:

   (a) $x^3 - x^2 - 2x \equiv 0 \pmod 5$

   (b) $x^3 + x^2 - 2 \equiv 0 \pmod 5$

   (c) $x^2 + 1 \equiv 0 \pmod{13}$

4. Solve the congruences in problem 3 for the second and third powers of the moduls.

CHAPTER 11

WILSON'S THEOREM AND ITS CONSEQUENCES

11-1. Wilson's theorem. In the *Meditationes Algebraicae* by Edward Waring, published in Cambridge in 1770, one finds, as we have already mentioned, several announcements on the theory of numbers. One of them is the following: For any prime p the quotient

$$\frac{1 \cdot 2 \cdot \cdots \cdot (p-1) + 1}{p}$$

is an integer.

This result Waring ascribes to one of his pupils John Wilson (1741-1793). Wilson was a senior wrangler at Cambridge and left the field of mathematics quite early to study law. Later he became a judge and was knighted. Waring gives no proof of Wilson's theorem until the third edition of his *Meditationes*, which appeared in 1782. Wilson probably arrived at the result through numerical computations. Among the posthumous papers of Leibniz there were later found similar calculations on the remainders of $n!$, and he seems to have made the same conjecture. The first proof of the theorem of Wilson was given by J. L. Lagrange in a treatise that appeared in 1770.

We shall prefer to give Wilson's theorem in the now usual congruence form:

THEOREM 11-1. For any prime p one has

$$(p - 1)! \equiv -1 \pmod{p} \tag{11-1}$$

The theorem is easily verified for small values of p.

$$1! \equiv -1 \pmod 2, \qquad 2! \equiv -1 \pmod 3, \qquad 4! \equiv -1 \pmod 5$$

Before we proceed to the general proof we shall indicate its main idea in the special case $p = 19$. For this modulus one has the congruences

$$2 \cdot 10 \equiv 1, \qquad 7 \cdot 11 \equiv 1$$
$$3 \cdot 13 \equiv 1, \qquad 8 \cdot 12 \equiv 1$$
$$4 \cdot 5 \equiv 1, \qquad 9 \cdot 17 \equiv 1$$
$$6 \cdot 16 \equiv 1, \qquad 14 \cdot 15 \equiv 1$$

and also

$$1 \cdot 1 \equiv 1, \qquad 18 \cdot 18 \equiv 1$$

When the first group of congruences is multiplied together and the numbers rearranged, it follows that

$$2 \cdot 3 \cdot \cdots \cdot 16 \cdot 17 \equiv 1 \pmod{19}$$

This congruence is multiplied by

$$18 \equiv -1 \pmod{19}$$

and we obtain

$$18! \equiv -1 \pmod{19}$$

as required by Wilson's theorem.

The proof in the general case proceeds along the same lines. Let p be some prime and a one of the numbers

$$1, 2, \ldots, p - 1 \tag{11-2}$$

The linear congruence

$$ax \equiv 1 \pmod{p}$$

as we have seen, has a single solution

$$x \equiv b \pmod{p}$$

In this manner there corresponds to every a in (11–2) a unique b, such that

$$ab \equiv 1 \pmod{p}$$

and clearly b corresponds to a in the same way. This shows that the numbers in (11–2) can be divided into pairs a, b whose product is congruent to 1 (mod p). The two numbers in a pair x, x can only be equal when

$$x^2 \equiv 1 \pmod{p}$$

This can be written

$$(x - 1)(x + 1) \equiv 0 \pmod{p}$$

so that it can occur only when

$$x \equiv 1 \pmod{p}$$

or

$$x \equiv -1 \equiv p - 1 \pmod{p}$$

that is, when $x = 1$ or $x = p - 1$. In multiplying the numbers (11–2) together to form $(p - 1)!$, the pairs with different a and b will give products congruent to 1 (mod p), while the two remaining factors 1 and $p - 1$ have a product that is congruent to -1. This completes the proof of Wilson's congruence (11–1).

One may ask what happens for other moduls. One computes easily

$$1! \equiv -1 \pmod{2}, \qquad 6! \equiv -1 \pmod{7}$$

$$2! \equiv -1 \pmod{3}, \qquad 7! \equiv 0 \pmod{8}$$

$$3! \equiv 2 \pmod{4}, \qquad 8! \equiv 0 \pmod{9}$$

$$4! \equiv -1 \pmod{5}, \qquad 9! \equiv 0 \pmod{10}$$

$$5! \equiv 0 \pmod{6}, \qquad 10! \equiv -1 \pmod{11}$$

The general result is:

THEOREM 11–2. For a composite number n one has

$$(n - 1)! \equiv 0 \pmod{n}$$

except when $n = 4$.

The proof is quite simple. Let us write

$$n = pq$$

where p is a prime. If p is not equal to q, both p and q occur as factors in the product $(n - 1)!$ so that it is divisible by n. When

$p = q$ and $n = p^2$, the factors p and $2p$ occur in $(n - 1)!$, provided $p > 2$. In the remaining case when $p = 2$, $n = 4$, there is an exception to the rule, as we found above.

Theorem 11–2 shows that Wilson's congruence (11–1) will hold for the primes and for no other numbers.

We shall deduce certain consequences from Wilson's congruence. In the product

$$(p - 1)! = 1 \cdot 2 \cdot \cdots \cdot \frac{p - 1}{2} \cdot \frac{p + 1}{2} \cdot \cdots \cdot (p - 2)(p - 1)$$

one has the following congruences:

$$p - 1 \equiv -1, \qquad p - 2 \equiv -2, \cdots,$$
$$\frac{p + 1}{2} \equiv -\frac{p - 1}{2} \qquad (\bmod\ p)$$

for the series of factors. Consequently

$$(p - 1)! \equiv (-1)^{\frac{p-1}{2}} \left(1 \cdot 2 \cdot \cdots \cdot \frac{p - 1}{2}\right)^2 \ (\bmod\ p)$$

and when this is substituted in Wilson's congruence, one finds

$$\left(1 \cdot 2 \cdot \cdots \cdot \frac{p - 1}{2}\right)^2 \equiv (-1)^{\frac{p+1}{2}} \ (\bmod\ p) \qquad (11\text{–}3)$$

Let us discuss this result a little further. When p is a prime of the form $4n + 1$,

$$(-1)^{\frac{p+1}{2}} = (-1)^{2n+1} = -1$$

so that (11–3) takes the form

$$\left(1 \cdot 2 \cdot \cdots \cdot \frac{p - 1}{2}\right)^2 + 1 \equiv 0 \ (\bmod\ p)$$

and we can state:

THEOREM 11–3. When p is a prime of the form $4n + 1$, the congruence

$$x^2 + 1 \equiv 0 \ (\bmod\ p)$$

is solvable and has the roots

$$x \equiv \pm \left(\frac{p-1}{2}\right)! \pmod{p}$$

For instance when $p = 13$ we find

$$6! \equiv 5 \pmod{13}$$

and

$$5^2 + 1 \equiv 0 \pmod{13}$$

In the second case when p is a prime of the form $4n + 3$, the congruence (11–3) reduces to

$$\left(1 \cdot 2 \cdot \cdots \cdot \frac{p-1}{2}\right)^2 - 1 \equiv 0 \pmod{p}$$

This may be written

$$\left[\left(\frac{p-1}{2}\right)! + 1\right]\left[\left(\frac{p-1}{2}\right)! - 1\right] \equiv 0 \pmod{p}$$

and one concludes:

THEOREM 11–4. For any prime p of the form $4n + 3$, one has one of the congruences

$$\left(\frac{p-1}{2}\right)! \equiv \pm 1 \pmod{p} \tag{11–4}$$

One finds for the lowest primes

$$1! \equiv 1 \pmod{3}, \qquad 3! \equiv -1 \pmod{7}, \qquad 5! \equiv -1 \pmod{11}$$

$$9! \equiv -1 \pmod{19}, \qquad 11! \equiv 1 \pmod{23}$$

There exists a complicated rule determining whether one shall use $+1$ or -1 in the congruence (11–4).

11–2. Gauss's generalization of Wilson's theorem. In the third section of the *Disquisitiones* Gauss indicates, without giving the details of the proof, how the theorem of Wilson can be extended to arbitrary moduls. Before we proceed to this theorem, it is necessary to carry through a certain auxiliary investigation.

We wish to determine the number of solutions of the congruence

$$x^2 \equiv 1 \pmod{m} \tag{11-5}$$

for a given modul m. As we have already observed previously, we can solve the congruence first for prime-power moduls and then obtain the solution for a general modul by the Chinese remainder theorem. Therefore, let p be a prime and let us study the congruence

$$x^2 \equiv 1 \pmod{p^\alpha} \tag{11-6}$$

This may be written

$$(x - 1)(x + 1) \equiv 0 \pmod{p^\alpha} \tag{11-7}$$

If $p > 2$, only one of these factors can be divisible by p so that one has either

$$x \equiv 1 \pmod{p^\alpha}$$

or

$$x \equiv -1 \pmod{p^\alpha}$$

and we conclude:

For a prime $p > 2$, the congruence (11-6) has two solutions

$$x \equiv \pm 1 \pmod{p^\alpha}$$

The case when $p = 2$ is slightly more complicated. For $\alpha = 1$ in (11-6), the congruence becomes

$$x^2 \equiv 1 \pmod{2}$$

which has a single solution $x \equiv 1 \pmod{2}$. For $\alpha = 2$, the congruence

$$x^2 \equiv 1 \pmod{4}$$

has two solutions $x \equiv \pm 1 \pmod{4}$. Finally let $\alpha > 2$. If one of the factors in (11-7) is divisible by 2, so is the other, but only one of them can be divisible by 4 'or a higher power of 2. If $x + 1$ is divisible only by 2 to the first power, one must have

$$x \equiv 1 \pmod{2^{\alpha-1}}$$

and this represents two different solutions

$$x \equiv 1, \qquad x \equiv 1 + 2^{\alpha-1} \qquad (\text{mod } 2^\alpha)$$

of the original congruence. Similarly when $x - 1$ contains only the first power of 2, one finds the solutions

$$x \equiv -1, \qquad x \equiv -1 - 2^{\alpha-1} \qquad (\text{mod } 2^\alpha)$$

The four solutions are different (mod 2^α), as one easily checks. To sum up, for $p = 2$, the congruence (11–6) has four solutions

$$x \equiv \pm 1, \qquad x \equiv \pm(1 + 2^{\alpha-1}) \qquad (\text{mod } 2^\alpha)$$

except when $\alpha = 2$, when there are two solutions

$$x \equiv \pm 1 \ (\text{mod } 4)$$

and when $\alpha = 1$, when there is a single solution

$$x \equiv 1 \ (\text{mod } 2)$$

It remains to determine the number of solutions of the general congruence (11–5). We decompose the modul into its prime factors

$$m = 2^\alpha p_1^{\beta_1} \cdots p_r^{\beta_r} \tag{11–8}$$

and solve the congruence (11–6) for the moduls 2^α and $p_i^{\beta_i}$. The Chinese remainder theorem shows that the solutions of (11–5) are obtained by selecting a particular solution for each of the prime powers and combining them by simultaneous congruences. When m is not divisible by 2, hence when $\alpha = 0$, each of the congruences (mod $p_i^{\beta_i}$) has 2 roots so that we obtain a total of 2^r solutions. When $\alpha = 1$ the congruence (mod 2) has a single solution so that there will still be 2^r solutions. When $\alpha = 2$ the congruence (mod 4) has two solutions, and the total number in this case is 2^{r+1}. Finally, when $\alpha > 2$, the congruence (mod 2^α) has four solutions so that the total number of solutions of (11–5) is 2^{r+2}. Thus we can state:

THEOREM 11–5. The congruence

$$x^2 \equiv 1 \ (\text{mod } m)$$

where the modul has the prime decomposition

$$m = 2^\alpha p_1{}^{\beta_1} \cdots p_r{}^{\beta_r}$$

will have

2^r solutions when $\alpha = 0$ or $\alpha = 1$

2^{r+1} solutions when $\alpha = 2$

2^{r+2} solutions when $\alpha > 2$

This completes our preparations for Gauss's generalization of Wilson's theorem:

THEOREM 11–6.   If one forms the product P of the remainders relatively prime to the number m, then

$$P \equiv \pm 1 \pmod{m} \tag{11–9}$$

In this congruence one has the value $+1$ in all cases, with the following exceptions where -1 appears:

1. $m = 4$.

2. $m = p^\beta$ is a power of an odd prime.

3. $m = 2p^\beta$ is twice the power of an odd prime.

When $m = 2$ it is immaterial whether one uses $+1$ or -1.

The result, as one sees, implies Wilson's theorem.   For the smallest composite moduls one finds

$$1 \cdot 3 \equiv -1 \pmod 4, \qquad 1 \cdot 2 \cdot 4 \cdot 5 \cdot 7 \cdot 8 \equiv -1 \pmod 9$$

$$1 \cdot 5 \equiv -1 \pmod 6, \qquad 1 \cdot 3 \cdot 7 \cdot 9 \equiv -1 \pmod{10}$$

$$1 \cdot 3 \cdot 5 \cdot 7 \equiv 1 \pmod 8, \qquad 1 \cdot 5 \cdot 7 \cdot 11 \equiv 1 \pmod{12}$$

The proof is based on the same principle as our proof of Wilson's theorem.   The set of all positive integers less than and relatively prime to m are paired as before.   To each such number a one can find a unique b for which

$$ab \equiv 1 \pmod{m}$$

Let us suppose first that a and b are different. Then in computing the remainder of the product P in (11–9), they may be disregarded since their product is congruent to 1 (mod m). In P, therefore, we need only to consider the product of those numbers a that belong to a pair a, a with equal components, *i.e.*, the solutions of the congruence

$$x^2 \equiv 1 \ (\text{mod } m) \tag{11–10}$$

If a is a solution of this congruence, so is $-a$ and these two numbers represent different remainders (mod m) since

$$a \equiv -a \ (\text{mod } m)$$

can only occur in the trivial case $m = 2$. In forming the remainder of the product P, we multiply a and $-a$ and note that

$$a(-a) = -a^2 \equiv -1 \ (\text{mod } m)$$

Each pair of roots a and $-a$ therefore contributes a factor -1 to the remainder of P (mod m) and the congruence (11–9) follows. The remainder $+1$ must be used when there is an even set of roots a and $-a$, hence when the number of roots of (11–10) is divisible by 4; otherwise one must use -1. But in theorem 11–5 this number of roots has been determined, and one verifies that the only cases in which it is not divisible by 4 are exactly those that have been enumerated above in Gauss's theorem.

We notice further, according to theorem 11–5, that the integers m for which the negative sign must be used in (11–9) are those for which the congruence

$$x^2 \equiv 1 \ (\text{mod } m)$$

has only two solutions and clearly these must be

$$x \equiv \pm 1 \ (\text{mod } m)$$

Problems.

Check the theorem of Gauss for the composite numbers below 25.

11–3. Representations of numbers as the sum of two squares. We have already mentioned certain results regarding the representation of numbers as the sum of two squares, both in discussing

Fermat's notes on Diophantos and in connection with the factorization of numbers. We are now ready to give the proofs for some of the basic facts. There are numerous ways in which this theory may be treated. We shall prefer a method based on a simple theorem on congruences given by the Norwegian mathematician Axel Thue (1863–1922), who is known for his important contributions to the newer theory of Diophantine equations.

The theorem of Thue is of interest also because it gives a simple application of a mathematical method known as *Dirichlet's box principle:* If one has n boxes and more than n objects to distribute in them, at least one of the boxes must contain more than one object.

This statement sounds extremely trivial, but, nevertheless, it has important applications to various questions in number theory.

The theorem of Thue that we wish to prove is the following:

THEOREM 11–7.   Let p be a prime and k the least integer greater than $\sqrt{p}$. Then for any integer a not divisible by p, one can find numbers x and y belonging to the set $1, 2, \ldots, k - 1$ such that

$$xa \equiv \pm y \pmod{p} \qquad (11\text{–}11)$$

Before we proceed to the proof, let us take as an example the case where $p = 23$ and $k = 5$. For $a = 9$ and $a = 10$ one finds, respectively,

$$3a \equiv 4, \qquad 2a \equiv -3 \pmod{23}$$

The reader may give a complete set of such congruences for all remainders $a \pmod{23}$.

To prove the theorem let us take all numbers

$$ax - y, \qquad x, y = 0, 1, \cdots, k - 1 \qquad (11\text{–}12)$$

and classify them according to their remainders $\pmod{p}$. There are altogether $k^2 > p$ such numbers (11–12), so that according to Dirichlet's box principle at least two of them must have the same remainder, hence be congruent $\pmod{p}$. Let us suppose that

$$ax_1 - y_1 \equiv ax_2 - y_2 \pmod{p}$$

or

$$a(x_1 - x_2) \equiv y_1 - y_2 \pmod{p} \qquad (11\text{–}13)$$

The absolute values of the differences

$$x_1 - x_2, \qquad y_1 - y_2$$

again belong to the set $0, 1, 2, \ldots, k - 1$. But the value 0 is excluded, because, for instance, $y_1 = y_2$ would involve, according to (11–13),

$$a(x_1 - x_2) \equiv 0 \pmod{p}$$

from which one concludes that $x_1 \equiv x_2$, consequently $x_1 = x_2$, contrary to the assumption that we are dealing with two different numbers in (11–12). In (11–13) we have, therefore, a congruence of the desired form (11–11) when one, if necessary, adjusts the sign so that the coefficient of a is positive.

In the next step we show:

THEOREM 11–8. A prime p is representable as the sum of two squares if the congruence

$$a^2 + 1 \equiv 0 \pmod{p} \tag{11–14}$$

is solvable.

To prove this result we take a solution a of the congruence (11–14) and determine the two numbers x and y according to theorem 11–7. We multiply the congruence (11–14) by x^2 and obtain

$$x^2 a^2 + x^2 \equiv y^2 + x^2 \equiv 0 \pmod{p}$$

so that

$$x^2 + y^2 = tp \tag{11–15}$$

for some suitable positive integer t. But x and y were chosen such that

$$x^2 \leqq (k - 1)^2 < p, \qquad y^2 \leqq (k - 1)^2 < p$$

and we conclude from (11–15)

$$pt < p + p = 2p$$

or $t < 2$. The only possibility is, therefore, $t = 1$ and

$$p = x^2 + y^2$$

is the desired representation of p as the sum of two squares.

When theorem 11–8 is combined with theorem 11–3 we arrive at the key result:

THEOREM 11–9.   Any prime of the form $4n + 1$ can be represented as the sum of two squares.

The prime 2 is evidently the sum of two squares, but no prime of the form $4n + 3$ can be the sum of two squares, because such a sum is never of the form $4n + 3$, a fact we have already mentioned in Chap. 4 in connection with the factorization method based on the representation of a number as a sum of two squares.   At that time we also showed that a prime can have but a single such representation, and we gave the representation of all primes below 100 as the sum of two squares.   Extensive tables of this kind have been computed.

THEOREM 11–10.   Let N be a positive integer and n^2 its greatest square factor so that

$$N = N_0 n^2$$

The necessary and sufficient condition for N to be representable as the sum of two squares is that N_0 contain no prime factors of the form $4n + 3$.

In Chap. 8–5 in discussing Fermat's notes on Diophantos's *Arithmetics*, we proved that on the basis of theorem 11–9 it follows that N can be represented as the sum of two squares provided the prime factors of N_0 were 2 or of the form $4n + 1$.

There remains, after what we have just said, only to prove that if N_0 has a prime factor $p_0 = 4n + 3$, no representation of N as the sum of two squares can exist.   This we achieve by showing that a decomposition

$$N = x^2 + y^2 \tag{11–16}$$

must lead to a contradiction.   Let us suppose that in (11–16) the greatest common divisor of x and y is d so that

$$x = dx_1, \qquad y = dy_1$$

We divide (11–16) by d^2 and find

$$\frac{N}{d^2} = \frac{n^2}{d^2} N_0 = x_1{}^2 + y_1{}^2 \tag{11–17}$$

where d^2 must divide n^2, and x_1 and y_1 are relatively prime. Since N_0 is divisible by p_0, we obtain from (11–17) the congruence

$$x_1{}^2 + y_1{}^2 \equiv 0 \pmod{p_0} \qquad (11\text{–}18)$$

One of the numbers x_1 and y_1, for instance x_1, is not divisible by p_0, so that we can find some z such that

$$x_1 z \equiv 1 \pmod{p_0}$$

When (11–18) is multiplied by z^2, it follows that

$$(y_1 z)^2 + 1 \equiv 0 \pmod{p_0}$$

This, however, according to theorem 11–8, would show that p_0 must be the sum of two squares, notwithstanding the fact that it is a prime of the form $4n + 3$. We have therefore established the impossibility of a representation (11–16).

Problem. Represent all primes of the form $4n + 1$ between 500 and 600 as the sum of two squares.

CHAPTER 12

EULER'S THEOREM AND ITS CONSEQUENCES

12-1. Euler's theorem. In a letter to Frénicle de Bessy dated October 18, 1640, Fermat writes as follows:

It seems to me after this that I should tell you the foundation on which I support the demonstrations of all which concerns geometric progressions, namely:

Every prime number measures [divides] infallibly one of the powers minus unity in any progression, and the exponent of this power is a divisor of the given prime number minus one; and after one has found the first power which satisfies the condition, all those whose exponents are multiples of the first satisfy the condition.

Fermat uses the example of the powers of three

1,	2,	3,	4,	5,	6
3,	9,	27,	81,	243,	729

where the first line gives the exponents. He points out that $3^3 - 1$ is the first such expression that is divisible by 13 and that the exponent 3 divides $13 - 1 = 12$ so that $3^{12} - 1$ is divisible by 13.

Fermat continues: "And this proposition is generally true for all series and all prime numbers. I would send you the demonstration, if I did not fear it being too long."

Unfortunately Fermat's correspondents never seemed particularly interested in demanding information about proofs for his results.

In congruence terminology Fermat states that for any number a and any prime p, there exists some smallest exponent d such that

$$a^d - 1 \equiv 0 \pmod{p}$$

272

and d divides $p - 1$, hence

$$a^{p-1} - 1 \equiv 0 \pmod{p}$$

It should be observed that in this theorem one must place the obvious restriction that a shall not be divisible by p.

For this result, as well as for several other of Fermat's theorems, Euler was the first to publish a proof; it appeared in the *Proceedings* of the St. Petersburg Academy in 1736. Considerably later (1760), Euler gave a more general theorem of the same kind, which we shall prefer to deduce first and then let Fermat's theorem follow as a special case.

To formulate Euler's theorem we first recall the definition of Euler's φ-function, which we studied in Chap. 5. For any positive integer m we denoted by $\varphi(m)$ the number of remainders from 1 to m that were relatively prime to m. For this function of m, we found the expression

$$\varphi(m) = m \left(1 - \frac{1}{p_1}\right) \cdots \left(1 - \frac{1}{p_r}\right) \tag{12-1}$$

where $p_1, p_2, \ldots, p_r$ are the *different* primes dividing m. Euler's theorem is then:

THEOREM 12–1. For any number a that is relatively prime to m one has the congruence

$$a^{\varphi(m)} \equiv 1 \pmod{m} \tag{12-2}$$

Before we proceed to the proof let us consider some examples.

Examples.

1. Let $m = 60$ and $a = 23$. One finds

$$\varphi(60) = 60(1 - \tfrac{1}{2})(1 - \tfrac{1}{3})(1 - \tfrac{1}{5}) = 16$$

and to verify the congruence (12–2) we compute for the modulus 60

$$23^2 \equiv -11$$

$$23^4 \equiv 121 \equiv 1$$

$$23^{16} \equiv 1$$

2. Let us take $m = 11$ and $a = 2$, hence

$$\varphi(11) = 11(1 - \tfrac{1}{11}) = 10$$

We have

$$2^5 = 32 \equiv -1$$

$$2^{10} \equiv 1 \qquad (\text{mod } 11)$$

To prove Euler's congruence (12-2) we denote the $\varphi(m)$ remainders less than and relatively prime to m by

$$r_1, r_2, \ldots, r_{\varphi(m)} \tag{12-3}$$

We multiply each of them by the given number a relatively prime to m and divide each product by m:

$$r_i a = q_i m + r_i' \tag{12-4}$$

where r_i' is the least positive remainder. Here r_i' must be relatively prime to m because in (12-4) a common factor of m and r_i' would divide $r_i a$ and this is impossible, since r_i and a are both relatively prime to m. The number r_i' is, therefore, also one of the remainders (12-3).

We shall prefer to write the relations (12-4) as congruences

$$r_i a \equiv r_i' \ (\text{mod } m) \tag{12-5}$$

Two different remainders r_i and r_j in (12-3) cannot give rise to the same r_i' in (12-5) because the congruence

$$r_i a \equiv r_j a \ (\text{mod } m)$$

implies, since a may be canceled, that

$$r_i \equiv r_j \ (\text{mod } m)$$

or $r_i = r_j$. We conclude that in (12-5) the remainders r_i and r_i' both run through the whole set (12-3).

Let us illustrate the situation on the example where $m = 20$ and $a = 7$. Here

$$\varphi(20) = 20(1 - \tfrac{1}{2})(1 - \tfrac{1}{5}) = 8$$

and the relatively prime residues are the eight numbers

$$1, 3, 7, 9, 11, 13, 17, 19$$

One finds

$$
\begin{array}{ll}
1a \equiv 7, & 11a \equiv 17 \\
3a \equiv 1, & 13a \equiv 11 \\
7a \equiv 9, & 17a \equiv 19 \\
9a \equiv 3, & 19a \equiv 13
\end{array}
\quad \text{(mod 20)}
$$

When all these congruences are multiplied together, one obtains

$$a^8 \cdot 1 \cdot 3 \cdot 7 \cdot 9 \cdot 11 \cdot 13 \cdot 17 \cdot 19$$
$$\equiv 1 \cdot 3 \cdot 7 \cdot 9 \cdot 11 \cdot 13 \cdot 17 \cdot 19 \quad \text{(mod 20)}$$

The product of the remainders is relatively prime to the modul and so may be canceled to give

$$a^8 \equiv 1 \ \text{(mod 20)}$$

The general proof is quite analogous. We multiply all $\varphi(m)$ congruences (12–5) and find

$$a^{\varphi(m)} r_1 r_2 \cdots r_{\varphi(m)} \equiv r_1' r_2' \cdots r_{\varphi(m)}' \ \text{(mod } m) \qquad \text{(12–6)}$$

Since the numbers r_i and r_i' form the set (12–3) of remainders, their products are equal. Furthermore, the r_i's are relatively prime to m so that in (12–6) the product can be canceled and there remains Euler's congruence (12–2).

The subsequent sections are all based on Euler's congruence. At this point we shall mention only one minor application to our familiar problem of solving linear congruences, or equivalently, linear indeterminate equations.

THEOREM 12–2. When a and m are relatively prime, the solution of the linear congruence

$$ax \equiv b \ \text{(mod } m) \qquad \text{(12–7)}$$

is given by the formula

$$x \equiv ba^{\varphi(m)-1} \ \text{(mod } m) \qquad \text{(12–8)}$$

For the proof we multiply both sides of the congruence (12–7) by $a^{\varphi(m)-1}$ and find from Euler's congruence

$$a^{\varphi(m)}x \equiv x \equiv ba^{\varphi(m)-1} \pmod{m}$$

Examples.

1 In the congruence

$$7x \equiv 5 \pmod{24}$$

we have $\varphi(24) = 8$ and therefore according to (12–8)

$$x \equiv 5 \cdot 7^7 \pmod{24}$$

To compute the smallest remainder of x we notice that

$$7^2 \equiv 1, \qquad 7^6 \equiv 1, \qquad 7^7 \equiv 7 \qquad \pmod{24}$$

and

$$x \equiv 5 \cdot 7 \equiv 11 \pmod{24}$$

2. Let us take the congruence

$$11x \equiv 9 \pmod{29}$$

Here $\varphi(29) = 28$ so that

$$x \equiv 9 \cdot 11^{27} \pmod{29}$$

One computes

$$11^2 \equiv 5$$
$$11^4 \equiv 25 \equiv -4$$
$$11^8 \equiv 16 \qquad\qquad\qquad \pmod{29}$$
$$11^{16} \equiv 256 \equiv -5$$
$$11^{27} = 11^{16} \cdot 11^8 \cdot 11^2 \cdot 11 \equiv (-5) \cdot 16 \cdot 5 \cdot 11 \equiv 8$$

and

$$x \equiv 9 \cdot 8 \equiv 14 \pmod{29}$$

The formula (12–8) is interesting because it gives the solution of the linear congruence in explicit form. However, for congruences that involve fairly large numbers, one finds that in regard to simplicity of computations it is definitely inferior to our previous method based on the linear indeterminate equations.

Problems.

1. Verify Euler's congruence in the following examples:

$$m = 81, \qquad a = 5$$

$$m = 120, \qquad a = 7, \qquad a = 19$$

$$m = 59, \qquad a = 2$$

2. Solve the following congruences by Euler's theorem:

$$7x \equiv 2 \pmod{24}$$

$$4x \equiv 3 \pmod{49}$$

$$17x \equiv 41 \pmod{620}$$

3. Compare the work involved by solving the congruence

$$311x \equiv 19 \pmod{203}$$

by Euler's theorem and by linear indeterminate equations.

12–2. Fermat's theorem. The theorem of Euler will now be applied to certain important special cases. We assume first that the modul $m = p^\alpha$ is a power of a prime. Then

$$\varphi(m) = p^\alpha \left(1 - \frac{1}{p} \right) = p^\alpha - p^{\alpha-1}$$

and the numbers relatively prime to m are those that are not divisible by p. Euler's theorem states therefore:

THEOREM 12–3. If the number a is not divisible by the prime p, one has

$$a^{p^\alpha - p^{\alpha-1}} \equiv 1 \pmod{p^\alpha} \tag{12–9}$$

By specializing further to the case where the modul $m = p$ is a prime, we arrive at Fermat's original theorem:

THEOREM 12–4. When p is a prime and a some number not divisible by p, then

$$a^{p-1} \equiv 1 \pmod{p} \tag{12–10}$$

As an example, let us take $p = 101$ and $a = 2$. One computes

$$2^4 \equiv 16$$
$$2^8 \equiv 256 \equiv 54$$
$$2^{16} \equiv (54)^2 \equiv -13 \quad (\text{mod } 101)$$
$$2^{32} \equiv 169 \equiv -33$$
$$2^{64} \equiv (-33)^2 \equiv -22$$

and

$$2^{100} = 2^{64} \cdot 2^{32} \cdot 2^4 \equiv (-22)(-33)16 \equiv 1 \quad (\text{mod } 101)$$

According to Fermat's theorem the algebraic congruence

$$x^{p-1} - 1 \equiv 0 \quad (\text{mod } p)$$

has the $p - 1$ different solutions $1, 2, \ldots, p - 1$. When theorem 10–7 on algebraic congruences is applied to this case, it leads immediately to

THEOREM 12–5. For any prime modulus p, one has

$$x^{p-1} - 1 \equiv (x - 1)(x - 2) \cdots (x - (p - 1)) \pmod{p} \quad (12\text{–}11)$$

For instance when $p = 5$

$$(x - 1)(x - 2)(x - 3)(x - 4)$$
$$\equiv (x - 1)(x - 2)(x + 2)(x + 1)$$
$$= (x^2 - 1)(x^2 - 4) \equiv x^4 - 1 \quad (\text{mod } 5)$$

When one compares the coefficients of the powers of x on both sides of the congruence (12–11), one obtains various congruences relating to the numbers $1, 2, \ldots, p - 1$. If we take the constant term in which x does not occur, one has on the left the number -1 and on the right the product

$$(-1)(-2) \cdots [-(p - 1)] = (-1)^{p-1}(p - 1)!$$

so that

$$(-1)^{p-1}(p - 1)! \equiv -1 \quad (\text{mod } p)$$

This, as one sees, is the same as Wilson's congruence.

Fermat's congruence (12–10) holds for all numbers a except those divisible by p. However, it is possible to formulate this result in such a manner that it is valid for *every* number without exception. When the congruence (12–10) is multiplied by a, one finds

$$a^p \equiv a \pmod{p}$$

for every a not divisible by p. Clearly this congruence holds also for those a that are divisible by p, since then both sides are divisible by p. Therefore, we can restate Fermat's theorem as follows:

THEOREM 12–6. For a prime p one has

$$a^p \equiv a \pmod{p} \tag{12–12}$$

for any a.

As a minor application of this theorem let us show:

THEOREM 12–7. In the decadic system a number and its fifth power have the same final digit.

In terms of congruences we shall have to establish that

$$a^5 \equiv a \pmod{10}$$

for any a. But by Fermat's congruence one has

$$a^5 \equiv a \pmod{5}$$

and trivially one finds

$$a^5 \equiv a \pmod{2}$$

Problems.

1. Verify Fermat's congruence (12–10) for

$$p = 71, \ a = 3, \qquad p = 59, \ a = 2$$

2. Verify the congruence (12–11) for $p = 11$ and $p = 13$.

3. Show that for any number n not divisible by 2 and 3 one has

$$n^2 \equiv 1 \pmod{24}$$

12–3. Exponents of numbers. Euler's theorem shows that to any number a that is relatively prime to the modul m there must exist some exponent n such that

$$a^n \equiv 1 \pmod{m} \tag{12–13}$$

The least positive exponent n for which this congruence (12–13) holds we shall call the *exponent to which a belongs* (mod m). For small moduls it may be determined directly by finding the residue (mod m) of the various powers of a.

For instance, let $m = 30$ and $a = 7$. One finds successively

$$a^2 \equiv -11, \qquad a^3 \equiv 13, \qquad a^4 \equiv 1 \qquad (\text{mod } 30)$$

so that 7 belongs to the exponent 4 (mod 30).

In connection with the initial statement of his theorem, which we quoted, Fermat points out in the example that the number 3 belongs to the exponent 3 (mod 13) and the exponent 3 divides $p - 1 = 12$.

In the study of the properties of the exponents to which a number belongs, we begin by mentioning:

THEOREM 12–8.   If the number a belongs to the exponent n (mod m) and if N is some other number such that

$$a^N \equiv 1 \ (\text{mod } m) \tag{12–14}$$

then n divides N.

We divide N by n

$$N = qn + r, \qquad 0 \leqq r < n$$

and deduce from (12–13) and (12–14)

$$a^N \equiv a^{qn+r} = (a^n)^q a^r \equiv a^r \equiv 1 \ (\text{mod } n)$$

Since n was the smallest positive exponent such that the congruence (12–13) would hold, we conclude that $r = 0$ and n divides N.

When theorem 12–8 is applied to Euler's congruence, there follows immediately:

THEOREM 12–9.   The exponent n to which a number a belongs (mod m) divides $\varphi(m)$.

This theorem brings up the important question: When a divisor n of $\varphi(m)$ is given, does there exist some number a belonging to the exponent n (mod m)?

In general this is not true, as one can show by examples. For instance, let us take $m = 15$ and $\varphi(m) = 8$. The numbers that are relatively prime to m are congruent to one of the eight numbers

$$\pm 1, \ \pm 2, \ \pm 4, \ \pm 7$$

One finds that all of them satisfy the congruence

$$x^4 \equiv 1 \ (\text{mod } 15)$$

so that there is no number belonging to the exponent $n = 8$.

On the other hand, let us compute the exponents to which the various numbers belong (mod 13). Here $\varphi(13) = 12$ and this number has the divisors

$$1, \ 2, \ 3, \ 4, \ 6, \ 12$$

One verifies that there are numbers belonging to each of these exponents (mod 13), namely:

Exponent	Numbers Belonging
1	1
2	12
3	3, 9
4	5, 8
6	4, 10
12	2, 6, 7, 11

This table illustrates the general result:

THEOREM 12–10. For a prime modul p there exist numbers belonging to every divisor n of $p - 1$.

The proof will be based on the following theorem, which throws light on the interrelation between the numbers that belong to the same exponent.

THEOREM 12–11. Let a be a number belonging to the exponent n for the prime modul p. Then the powers

$$a^{r_1}, \ a^{r_2}, \ \ldots, \ a^{r_{\varphi(n)}} \tag{12--15}$$

represent all numbers belonging to the same exponent where

$$r_1 = 1, \ r_2, \ \cdots, \ r_{\varphi(n)} \tag{12-16}$$

are the positive remainders less than and relatively prime to n.

For instance, in the example above, the number 2 belongs to the exponent $n = 12 \pmod{13}$. The remainders less than and relatively prime to 12 are

$$1, 5, 7, 11$$

and one finds as in the table that

$$2, \qquad 2^5 \equiv 6, \qquad 2^7 \equiv 11, \qquad 2^{11} \equiv 7, \qquad \pmod{13}$$

are the numbers belonging to the exponent 12 (mod 13).

To prove theorem 12–11 we notice first that a satisfies the congruence

$$x^n \equiv 1 \pmod{p} \tag{12–17}$$

But for any i one has

$$(a^i)^n = (a^n)^i \equiv 1 \pmod{p}$$

so that the n numbers

$$1, a, a^2, \ldots, a^{n-1} \tag{12–18}$$

satisfy the same congruence (12–17). Furthermore the n powers (12–18) are incongruent (mod p) because from a congruence

$$a^i \equiv a^j \pmod{p}$$

would follow

$$a^{i-j} \equiv 1 \pmod{p}$$

with a smaller exponent $i - j$ than n. We conclude therefore from the theorem of Lagrange (theorem 10–8) that the numbers (12–18) are all the roots of the congruence (12–17) of nth degree.

Consequently, the numbers belonging to the exponent $n \pmod{p}$ are to be found in (12–18). But all numbers $a_r = a^r$, where r has a common factor d with n, must belong to a smaller exponent since

$$a_r^{\frac{n}{d}} = (a^n)^{\frac{r}{d}} \equiv 1 \pmod{p}$$

As possible numbers belonging to the exponent n there are left only those $a_r = a^r$ where r is relatively prime to n, that is, where r is one of the remainders (12–16). Let n_r be the exponent to which a_r belongs (mod p). Since a_r is a root of the congruence (12–17), n_r must divide n. But conversely from

$$(a^r)^{n_r} \equiv 1 \pmod{p}$$

it follows that n divides $r \cdot n_r$, therefore, since $(r,n) = 1$, n also divides n_r, so that $n_r = n$. This completes the proof of theorem 12–11.

The deduction of theorem 12–10 from theorem 12–11 is based on the following reasoning. We denote by

$$n_1 = 1, \; n_2, \; n_3, \; \cdots, \; n_\nu = p - 1 \tag{12–19}$$

the set of all divisors of $p - 1$, and for any such divisor n_i let N_i be the number of integers belonging to the exponent n_i (mod p). Since any integer not divisible by p belongs to some exponent (mod p), we must have

$$N_1 + N_2 + N_3 + \cdots + N_\nu = p - 1 \tag{12–20}$$

Theorem 12–11 states that for any divisor n_i we have only two alternatives: either there is no integer belonging to this exponent (mod p) so that $N_i = 0$, or we have $N_i = \varphi(n_i)$ where φ is Euler's function. But in discussing Euler's φ-function we derived the theorem (theorem 5–12) that the sum of the φ-functions of the divisors of a number was equal to the number itself. When applied to the divisors (12–19) of $p - 1$, this gives

$$\varphi(n_1) + \varphi(n_2) + \cdots + \varphi(n_\nu) = p - 1 \tag{12–21}$$

By comparison of (12–20) and (12–21), one sees that both sums cannot have the same value $p - 1$ except when each N_i takes the value $\varphi(n_i)$. Thus for every divisor n_i of $p - 1$, there exist numbers belonging to it.

The proof shows that one can supplement theorem 12–10 as follows:

THEOREM 12–12. For every divisor n of $p - 1$, there exist $\varphi(n)$ numbers belonging to the exponent n for the prime modul p.

One may verify this result on the table of exponents (mod 13) given above.

Problems.

1. Find the exponents to which the relatively prime remainders belong for the following moduls:

$$m = 40, \quad m = 30, \quad m = 20, \quad m = 7$$

2. Check theorem 12–12 on the residues for the following primes:

$$p = 11, \quad p = 17, \quad p = 19$$

12–4. Primitive roots for primes. The highest possible exponent to which a number can belong (mod m) is $\varphi(m)$. If a number belongs to this maximal exponent, we shall call it a *primitive root* for the modul m. Not every modul has primitive roots, as we have already mentioned; for instance, for the modul $m = 15$, one finds that every remainder relatively prime to 15 will satisfy the congruence

$$x^4 \equiv 1 \pmod{15}$$

and yet $\varphi(15) = 8$. The determination of those moduls for which primitive roots can exist is one of the problems to be taken up in the following discussion.

From theorem 12–12 we conclude immediately:

THEOREM 12–13. For a prime modul p there exist $\varphi(p - 1)$ primitive roots.

From the previous table of the exponents to which the various remainders belong (mod 13), we see that

$$2, \ 6, \ 7, \ 11$$

are the primitive roots (mod 13).

To find the primitive roots of a modul if they exist, one must usually proceed by trial and error, although there are certain rules that may facilitate the search. Often one of the small numbers 2, 3, 5, or 6 may turn out to be a primitive root. The following table gives the smallest positive primitive root for all primes below 200.

Extensive tables of primitive roots for primes have been computed. The first of these, the *Canon arithmeticus* (1839) by K. G. J. Jacobi, included primitive roots for all primes below 1,000. More recent tables by Kraitchik, Cunningham, and others give primitive roots for all primes up to 25,000 and even beyond.

Prime	Primitive root	Prime	Primitive root
2	1	89	3
3	2	97	5
5	2	101	2
7	3	103	5
11	2	107	2
13	2	109	6
17	3	113	3
19	2	127	3
23	5	131	2
29	2	137	3
31	3	139	2
37	2	149	2
41	6	151	6
43	3	157	5
47	5	163	2
53	2	167	5
59	2	173	2
61	2	179	2
67	2	181	2
71	7	191	19
73	5	193	5
79	3	197	2
83	2	199	3

12–5. Primitive roots for powers of primes. We shall now tackle the question of finding all moduls for which there exist primitive roots. One of the main results in this direction is

THEOREM 12–14. There exist primitive roots for all powers of a prime $p > 2$.

The proof of this theorem is quite long, and we shall take it in several steps.   First we show:

There exist primitive roots when the modul is a square of a prime.

Since

$$\varphi(p^2) = p(p - 1)$$

every number a not divisible by p satisfies the congruence

$$a^{p(p-1)} \equiv 1 \pmod{p^2}$$

We shall take some primitive root r (mod p) and examine when it may be a primitive root (mod p^2).   Let r belong to the exponent d (mod p^2) so that

$$r^d \equiv 1 \pmod{p^2}$$

is the congruence with the smallest possible exponent that r satisfies (mod p^2).   Then d divides $p(p - 1)$.   On the other hand,

$$r^d \equiv 1 \pmod{p}$$

and r is a primitive root (mod p) so that d is a multiple of $p - 1$. The two possibilities for d are, therefore, either $d = p - 1$ or $d = p(p - 1)$.   In the latter case r is a primitive root (mod p^2). Thus we are concerned only with the alternative where

$$r^{p-1} \equiv 1 \pmod{p^2} \tag{12–22}$$

Clearly, when r satisfies this congruence (12–22) it is not a primitive root (mod p^2), but the situation may be remedied by using

$$r_1 = r + p$$

instead of r.   Since

$$r_1 \equiv r \pmod{p}$$

the new number is also a primitive root (mod p).   Furthermore, by the binomial expansion we have

$$r_1^{p-1} = r^{p-1} + \frac{p-1}{1} r^{p-2}p + \frac{(p-1)(p-2)}{1\cdot 2} r^{p-3}p^2 + \cdots$$

so that

$$r_1^{p-1} \equiv r^{p-1} + p(p-1)r^{p-2} \pmod{p^2}$$

From the congruence (12–22) follows further

$$r_1{}^{p-1} - 1 \equiv (p-1)pr^{p-2} \pmod{p^2}$$

This shows that r_1 does not satisfy the condition (12–22), consequently it is a primitive root (mod p^2).

We have established that a primitive root r (mod p^2) is a primitive root (mod p) such that the congruence (12–22) does not hold, that is,

$$r^{p-1} = 1 + tp \qquad (12\text{–}23)$$

where t is not divisible by p. We shall use this to show:

A number r with these properties is a primitive root for any power p^α.

For the proof it is necessary to establish the auxiliary fact that one always has

$$r^{p^{\alpha-1}\,(p-1)} = 1 + t_\alpha \cdot p^\alpha \qquad (12\text{–}24)$$

where the integer t_α is not divisible by p. We have assumed this to be true for $\alpha = 1$, and may therefore use induction to prove it in general. We raise both sides of (12–24) to the pth power and expand the right-hand side by the binomial theorem to obtain

$$r^{p^\alpha(p-1)} = 1 + \frac{p}{1} \cdot t_\alpha \cdot p^\alpha + \frac{p(p-1)}{1.2} \cdot t_\alpha{}^2 \cdot p^{2\alpha} + \cdots$$

where the subsequent terms contain p to powers with exponents at least equal to 3α. When we write this expression in the form

$$r^{p^\alpha(p-1)} = 1 + p^{\alpha+1}\left[t_\alpha + \frac{p-1}{2} \cdot t_\alpha{}^2 \cdot p^\alpha + \cdots\right]$$

we shall have to show that the number $t_{\alpha+1}$ represented by the bracket is not divisible by p. According to the induction assumption t_α is not divisible by p; the second term is divisible by p (it is integral since p is an odd prime; this is the only place in the proof of theorem 12–14 that this fact is used), and finally all terms in the bracket not written out explicitly are also divisible by p since $3\alpha > \alpha + 1$; thus t_α is not divisible by p as desired.

To prove that r is a primitive root for all powers of p we again proceed by induction and assume that r is a primitive root (mod p^α) and show that it is a primitive root (mod $p^{\alpha+1}$).

Let us suppose that r belongs to the exponent d (mod $p^{\alpha+1}$).

$$r^d \equiv 1 \pmod{p^{\alpha+1}}$$

We know that d is a divisor of

$$\varphi(p^{\alpha+1}) = p^\alpha(p-1)$$

But since

$$r^d \equiv 1 \pmod{p^\alpha}$$

and r is a primitive root (mod p^α), d is a multiple of $p^{\alpha-1}(p-1)$, so that only the two values

$$d = p^{\alpha-1}(p-1) = \varphi(p^\alpha), \qquad d = p^\alpha(p-1) = \varphi(p^{\alpha+1})$$

are possible for d. But the first of these is excluded by the condition (12–24), and r is a primitive root (mod $p^{\alpha+1}$).

As an example let us determine a primitive root for the powers of 7. One finds that 3 is a primitive root (mod 7) and since

$$3^6 - 1 \not\equiv 0 \pmod{7^2}$$

it is a primitive root (mod 7^2) and therefore for all higher powers of 7.

The powers of the prime 2 have been excluded from our considerations and here the situation is different. For the modul $p = 2$ there is the primitive root $r = 1$, and $r = 3$ (mod 2^2) is a primitive root. But for the modul 8 and for higher powers of 2, there are no primitive roots. The numbers relatively prime to the modulus are in this case the odd numbers

$$a = 4n \pm 1$$

The square is

$$a^2 = 1 \pm 8n + 16n^2$$

so that for all of them

$$a^2 \equiv 1 \pmod{8}$$

while $\varphi(8) = 4$. Again from

$$a^2 = 1 + 8i$$

one finds by successive squarings

$$a^4 \equiv 1 \pmod{16}$$

$$a^8 \equiv 1 \pmod{32}$$

and in general

$$a^{2^{\alpha-2}} \equiv 1 \pmod{2^\alpha} \tag{12-25}$$

Since

$$\varphi(2^\alpha) = 2^{\alpha-1}$$

the congruence (12–25) shows that there can be no primitive roots for the higher powers of 2.

The congruence (12–25) implies that the highest exponent to which a number can possibly belong (mod 2^α) is $2^{\alpha-2}$, when $\alpha \geq 3$. It is not difficult to find numbers belonging to this maximal exponent, for instance, $a = 3$ is one of them. This may be seen from the following sequence of congruences, where each is obtained from the preceding by squaring:

$$3 \equiv -1 + 4 \pmod{8}$$

$$3^2 \equiv 1 + 8 \pmod{16}$$

$$3^4 \equiv 1 + 16 \pmod{32}$$

$$\cdot \quad \cdot \quad \cdot \quad \cdot \quad \cdot \quad \cdot \quad \cdot \quad \cdot$$

$$3^{2^{\alpha-3}} \equiv 1 + 2^{\alpha-1} \pmod{2^\alpha}$$

Since $\varphi(2^\alpha)$ is a power of 2, every number belongs to an exponent that is also a power of 2. But our last congruence shows that when 3 is raised to the power $2^{\alpha-3}$ one still does not have the remainder 1 (mod 2^α). The power with the exponent $2^{\alpha-2}$ is therefore the first with this property. It may be noted that 3 is a primitive root (mod 2) and (mod 4) so that it belongs to the greatest possible exponent for any power of 2. Let us summarize the results:

THEOREM 12–15. Among the powers of 2 only 2 and 4 have primitive roots. For all higher powers of 2^α, $\alpha \geqq 3$, every odd number satisfies the congruence

$$a^{2^{\alpha-2}} = a^{\frac{1}{2}\varphi(2^\alpha)} \equiv 1 \pmod{2^\alpha}$$

The number 3 is a primitive root (mod 2) and (mod 4) and belongs to the highest possible exponent $2^{\alpha-2}$ when $\alpha \geqq 3$.

Problems.

1. Find primitive roots for the powers of all odd primes up to $p = 19$.
2. Find numbers other than 3 that belong to the exponent $2^{\alpha-2}$ (mod 2^α).

12–6. Universal exponents. After we have investigated the existence of primitive roots for powers of primes, it is relatively simple to find all numbers divisible by two or more different primes for which there can be primitive roots. Let

$$m = p^\alpha q^\beta \cdots \tag{12–26}$$

be some such number. For an arbitrary a relatively prime to m, we know by Euler's theorem that

$$a^{\varphi(p^\alpha)} \equiv 1 \pmod{p^\alpha} \tag{12–27}$$

for any prime power p^α occurring in (12–26). The least common multiple of the various exponents in these congruences (12–27) we shall denote by

$$M = [p^{\alpha-1}(p-1), \quad q^{\beta-1}(q-1), \cdots] \tag{12–28}$$

Then clearly also

$$a^M \equiv 1 \pmod{p^\alpha}$$

for every prime power p^α in (12–26), hence

$$a^M \equiv 1 \pmod{m} \tag{12–29}$$

for every a relatively prime to m.

If a primitive root is to exist (mod m), M cannot be less than $\varphi(m)$. But we know that

$$\varphi(m) = \varphi(p^\alpha) \, \varphi(q^\beta) \cdots$$

and this product can only be equal to the l.c.m. (12–28) of its factors when these are relatively prime. Ordinarily this is not the case since all numbers $\varphi(p^\alpha)$ are even, with the single exception $\varphi(2) = 1$. We conclude that only when m has the special form

$$m = 2p^\alpha$$

is there a possibility for a primitive root. But in this case such a root is easily obtained. We notice

$$\varphi(m) = \varphi(2) \cdot \varphi(p^\alpha) = \varphi(p^\alpha)$$

so that when r is a primitive root (mod p^α), we have

$$r^{\varphi(m)} \equiv 1 \pmod{p^\alpha} \tag{12–30}$$

and $\varphi(m)$ is the smallest exponent with this property. When r is odd, the congruence (12–30) holds also (mod 2), and therefore (mod m); consequently r is a primitive root (mod m). If r should happen to be even, it may be replaced by

$$r_1 = r + p^\alpha$$

which is an odd primitive root (mod p^α).

To recapitulate our various results on primitive roots we state:

THEOREM 12–16. There exist primitive roots only for the three following classes of moduls:

1. $m = p^\alpha$ is the power of an odd prime.
2. $m = 2p^\alpha$ is the double of the power of an odd prime.
3. $m = 2$ and $m = 4$.

The same class of numbers as those that have primitive roots we have encountered earlier in connection with Gauss's extension of Wilson's theorem.   On the basis of theorem 12-16 we can re-formulate Gauss's theorem as follows:

THEOREM 12-17.   The product P of all positive remainders less than and relatively prime to a number m satisfies the congruence

$$P \equiv \pm 1 \pmod{m}$$

where the sign -1 occurs if and only if there exists a primitive root (mod m).

According to a remark which we made in connection with theorem 11-6, we conclude also that a primitive root (mod m) can exist only if the congruence

$$x^2 \equiv 1 \pmod{m}$$

has no other roots than

$$x \equiv \pm 1 \pmod{m}$$

We shall mention another problem closely related to those we have discussed.   We know that for a given modul m there exist exponents M such that

$$a^M \equiv 1 \pmod{m}$$

for every a relatively prime to m; for instance $\varphi(m)$ is such an exponent according to Euler's congruence.   We may call such a number M a *universal exponent* (mod m), and among these there is a *minimal universal exponent* $\lambda(m)$.   This number $\lambda(m)$ is of importance in several questions in number theory, and we shall now show how it may be determined.

For a power p^α of an odd prime, one must have

$$\lambda(p^\alpha) = \varphi(p^\alpha)$$

because there exist primitive roots belonging to the exponent $\varphi(p^\alpha)$.

For the powers of 2 the expression for the minimal universal exponent is slightly more complicated.   Since there are primitive roots (mod 2) and (mod 4), one has

$$\lambda(2) = 1, \qquad \lambda(4) = 2$$

For the higher powers theorem 12–15 shows that

$$\lambda(2^\alpha) = 2^{\alpha-2} = \tfrac{1}{2}\varphi(2^\alpha), \qquad \alpha \geqq 3$$

If one prefers to include the three cases in a single statement, one can write

$$\lambda(2^\alpha) = 2^{\beta-2} \begin{cases} \beta = \alpha, \text{ when } \alpha \geqq 3 \\ \beta = 3, \text{ when } \alpha = 2 \\ \beta = 2, \text{ when } \alpha = 1 \end{cases} \qquad (12\text{–}31)$$

We now turn to the general case and prove:

THEOREM 12–18. Let m be an integer with the factorization

$$m = 2^{\alpha_0} \cdot p_1^{\alpha_1} \cdot p_2^{\alpha_2} \cdots \qquad (12\text{–}32)$$

into prime powers. The minimal universal exponent of m is the least common multiple

$$N = [\lambda(2^{\alpha_0}), \quad \varphi(p_1^{\alpha_1}), \quad \varphi(p_2^{\alpha_2}), \cdots] \qquad (12\text{–}33)$$

of the corresponding minimal exponents of the prime powers in (12–32). Furthermore, there exist numbers belonging to this exponent (mod m).

Since N, according to its definition, is divisible by the smallest universal exponent for each prime power p^α occurring in (12–32), we conclude that one has

$$a^N \equiv 1 \pmod{p^\alpha}$$

for any a not divisible by p, consequently also

$$a^N \equiv 1 \pmod{m} \qquad (12\text{–}34)$$

for any a relatively prime to m.

To complete the proof of theorem 12–18, it is sufficient to show that there exists some number a for which N as defined is the smallest exponent such that the congruence (12–34) holds. This may be achieved by taking a to satisfy the congruences

$$a \equiv 3 \pmod{2^{\alpha_0}}, \qquad a \equiv r_1 \pmod{p_1^{\alpha_1}},$$

$$a \equiv r_2 \pmod{p_2^{\alpha_2}}, \cdots \qquad (12\text{–}35)$$

where r_1, r_2, ... are primitive roots of the prime powers $p_1^{\alpha_1}$, $p_2^{\alpha_2}$, ... According to the Chinese remainder theorem, one can always determine some a fulfilling all conditions (12–35). For the various prime powers occurring in the decomposition (12–32) of m, the number a must belong respectively to the exponents

$$\lambda(2^{\alpha_0}), \quad \varphi(p_1^{\alpha_1}), \quad \varphi(p_2^{\alpha_2}), \quad \ldots$$

consequently the smallest exponent to which a belongs (mod m) is their l.c.m. $N = \lambda(m)$.

The number $\lambda(m)$, sometimes called the *indicator* of m, is a divisor of $\varphi(m)$, but $\lambda(m)$ may be considerably smaller than $\varphi(m)$ for composite m. For instance, let us take

$$m = 720 = 2^4 \cdot 3^2 \cdot 5$$

Here

$$\varphi(m) = 192$$

while

$$\lambda(m) = [2^2, \quad \varphi(9), \quad \varphi(5)] = [4, \quad 6, \quad 4] = 12$$

As a consequence every number not divisible by 2, 3, or 5 satisfies the congruence

$$a^{12} \equiv 1 \pmod{720}$$

We may notice that $\lambda(m)$ according to its definition is even when $m > 2$.

Problems.

1. Find the indicator $\lambda(m)$ for the following numbers:

$\qquad$ (a) $m = 365$ $\qquad$ (b) $m = 144$ $\qquad$ (c) $m = 5!$ $\qquad$ (d) $m = 10!$

In each case try to find a number belonging to the exponent $\lambda(m)$ (mod m).

2. Find a primitive root for the moduls

$\qquad$ (a) $m = 54$ $\qquad$ (b) $m = 50$ $\qquad$ (c) $m = 68$

12–7. Indices. We have solved the problem of finding all moduls for which there exist primitive roots. One of the reasons for placing emphasis on this question is that when such a root exists, one can introduce a curious theory reminiscent of logarithms, which in important cases facilitates the study of congruences.

Let r be a primitive root for some modul m. We consider the series

$$1, \quad r, \quad r^2, \quad \ldots, \quad r^{\varphi(m)-1} \qquad (12\text{--}36)$$

of the $\varphi(m)$ first powers of r. None of these can be congruent (mod m). Let us assume, for instance, that $i > j$ and

$$r^i \equiv r^j \ (\text{mod } m)$$

Since r is relatively prime to m, we can cancel a power of r and obtain

$$r^{i-j} \equiv 1 \ (\text{mod } m)$$

and this contradicts the fact that r as a primitive root belongs to the exponent $\varphi(m)$ (mod m).

The $\varphi(m)$ numbers (12–36) are all incongruent and relatively prime to m so that in some order they must be congruent to the $\varphi(m)$ relatively prime remainders (mod m). Therefore, for any number a relatively prime to m, one can find a unique exponent i such that

$$a \equiv r^i \ (\text{mod } m) \qquad (12\text{--}37)$$

where

$$0 \leq i \leq \varphi(m) - 1 \qquad (12\text{--}38)$$

This exponent i we shall call the *index of the number a for the root r* (mod m) and denote it by

$$i = \text{Ind}_r (a) \qquad (12\text{--}39)$$

Very often the root r remains the same throughout some discussion, and it may then be dropped in the notation (12–39). Among the special cases we notice particularly

$$\text{Ind}_r 1 = 0, \qquad \text{Ind}_r r = 1$$

As an example let us take the modul $m = 9$. Here $\varphi(m) = 6$ and $r = 2$ is a primitive root. We compute the various powers of r and find

$$
\begin{aligned}
&1, && 2^4 \equiv 7 \\
&2, && 2^5 \equiv 5 \qquad (\text{mod } 9) \\
&2^2 = 4, && 2^6 \equiv 1 \\
&2^3 = 8,
\end{aligned}
$$

This gives us the following table of the numbers with given indices (mod 9) for the root $r = 2$:

Index	0	1	2	3	4	5
Number	1	2	4	8	7	5

To obtain the index of a given number one rearranges the table to make the remainders relatively prime to 9 the primary entry.

Number	1	2	4	5	7	8
Index	0	1	2	5	4	3

In a similar example let us take $m = 23$. A primitive root is $r = 5$, and through a reduction of the powers of 5 (mod 23), one finds the table:

Index	0	1	2	3	4	5	6	7	8	9	10
Number	1	5	2	10	4	20	8	17	16	11	9
Index	11	12	13	14	15	16	17	18	19	20	21
Number	22	18	21	13	19	3	15	6	7	12	14

Through rearrangement of the entries in this table, one obtains the companion table:

Number	1	2	3	4	5	6	7	8	9	10	11
Index	0	2	16	4	1	18	19	6	10	3	9
Number	12	13	14	15	16	17	18	19	20	21	22
Index	20	14	21	17	8	7	12	15	5	13	11

In the definition (12–37) of the index, we assumed that it was limited to the range given in (12–38). It is practical to drop this convention and permit several indices for the same number. If for two numbers i_1 and i_2, we have simultaneously

$$a \equiv r^{i_1} \equiv r^{i_2} \pmod{m} \tag{12–40}$$

it follows that

$$r^{i_1 - i_2} \equiv 1 \pmod{m}$$

Since r belongs to the exponent $\varphi(m)$ (mod m) this congruence is possible only when $i_1 - i_2$ is a multiple of $\varphi(m)$, hence

$$i_1 \equiv i_2 \pmod{\varphi(m)} \tag{12–41}$$

Conversely, when the congruence (12–41) is fulfilled it is easily seen that (12–40) must hold. Therefore, in dealing with indices, two indices shall be considered to be the same when they are congruent (mod $\varphi(m)$) and relations between indices may be treated as congruences (mod $\varphi(m)$).

The defining congruence (12–37) for the index of a number can be written in the form

$$a \equiv r^{\mathrm{Ind}_r(a)} \pmod{m} \tag{12–42}$$

This is entirely analogous to the definition of *logarithms* to some base g by the common rule

$$a = g^{\log_g a}$$

and the indices could well have been called *congruence logarithms*. In regard to congruences, they have applications similar to those of the logarithms for real numbers. The idea of indices goes back to Euler, but Gauss gives the first systematic discussion in the third section of the *Disquisitiones*.

The basic law for logarithms is expressed in the formula for the logarithm of a product

$$\log (ab) = \log a + \log b$$

For indices one has analogously

$$\mathrm{Ind} (ab) \equiv \mathrm{Ind}\, a + \mathrm{Ind}\, b \pmod{\varphi(m)} \tag{12–43}$$

The proof is immediate: From

$$a \equiv r^{\text{Ind } a}, \qquad b \equiv r^{\text{Ind } b} \qquad (\text{mod } m)$$

one obtains by multiplication

$$a \cdot b \equiv r^{\text{Ind } (a \cdot b)} \equiv r^{\text{Ind } a + \text{Ind } b} \qquad (\text{mod } m)$$

from which the rule (12–43) follows.

Examples.

Let us take

$$m = 23, \qquad \varphi(m) = 22, \qquad a = 15, \qquad b = 11$$

so that

$$ab = 165 \equiv 4 \ (\text{mod } 23)$$

Here one finds from our previous table

$$\text{Ind } a = 17, \qquad \text{Ind } b = 9, \qquad \text{Ind } (ab) = 4$$

and this agrees with the rule (12–49) since

$$17 + 9 \equiv 4 \ (\text{mod } 22)$$

The formula (12–43) can be extended to an arbitrary number of factors. When applied to n equal factors a, one has the power rule

$$\text{Ind } a^n \equiv n \ \text{Ind } a \ (\text{mod } \varphi(m)) \qquad (12\text{–}44)$$

Among the applications of the index theory, let us consider first the solution of a linear congruence

$$ax \equiv b \ (\text{mod } m)$$

where a and b are relatively prime to the modul m. By taking indices, one finds according to (12–43)

$$\text{Ind } a + \text{Ind } x \equiv \text{Ind } b \ (\text{mod } \varphi(m))$$

or

$$\text{Ind } x \equiv \text{Ind } b - \text{Ind } a \ (\text{mod } \varphi(m))$$

Examples.

1. Let us solve the congruence

$$7x \equiv 2 \ (\text{mod } 9)$$

By means of the table of indices (mod 9), one concludes

$$\text{Ind } x \equiv \text{Ind } 2 - \text{Ind } 7 \equiv 1 - 4 \equiv 3 \pmod{6}$$

and the number corresponding to this index is

$$x \equiv 8 \pmod{9}$$

2. Next we solve the congruence

$$17x \equiv 9 \pmod{23}$$

by means of the table of indices (mod 23). One finds

$$\text{Ind } x \equiv \text{Ind } 9 - \text{Ind } 17 \equiv 10 - 7 = 3 \pmod{22}$$

and the number with this index is

$$x \equiv 10 \pmod{23}$$

It is seen from the examples that in the index calculus it is convenient to have a double set of tables, one with the number as entry giving the indices, and another with the indices as entries giving the corresponding numbers. In this respect the indices are not as easy to handle as the logarithms, since the values of the logarithms occur in order and one can use the same table to find both the logarithms and the antilogarithms.

Congruences of the type

$$ax^n \equiv b \pmod{m} \tag{12-45}$$

may be solved readily by means of indices. By the previous rules (12-43) and (12-44), one finds

$$\text{Ind } a + n \text{ Ind } x \equiv \text{Ind } b \pmod{\varphi(m)}$$

so that Ind x is the solution of the linear congruence

$$n \text{ Ind } x \equiv \text{Ind } b - \text{Ind } a \pmod{\varphi(m)} \tag{12-46}$$

Similarly, for an exponential congruence

$$ab^x \equiv c \pmod{m} \tag{12-47}$$

one finds the solution from

$$x \text{ Ind } b \equiv \text{Ind } c - \text{Ind } a \pmod{\varphi(m)} \tag{12-48}$$

It should be noted that the congruences (12-45) and (12-47) may have one or several or even no solutions, depending on the behavior of the resulting linear congruences (12–46) and (12–48) for the indices.

Examples.

1. Let us take
$$3x^5 \equiv 11 \pmod{23}$$
Here one finds
$$5 \text{ Ind } x \equiv \text{Ind } 11 - \text{Ind } 3 \pmod{22}$$
or
$$5 \text{ Ind } x \equiv -7 \pmod{22}$$
This congruence has a single solution
$$\text{Ind } x \equiv 3 \pmod{22}$$
and from the table one finds
$$x \equiv 10 \pmod{23}$$

2. We wish to find integers x such that
$$3x^{14} \equiv 2 \pmod{23}$$
This leads to
$$14 \text{ Ind } x \equiv \text{Ind } 2 - \text{Ind } 3 \equiv -14 \pmod{22}$$
and one finds two solutions
$$\text{Ind } x \equiv 10, \quad \text{Ind } x \equiv 21 \quad \pmod{22}$$
which correspond to the values
$$x \equiv 9, \quad x \equiv 14 \quad \pmod{23}$$

3. There is no solution to the congruence
$$13^x \equiv 5 \pmod{23}$$
because it leads to
$$x \text{ Ind } 13 \equiv \text{Ind } 5 \pmod{22}$$
or
$$14x \equiv 1 \pmod{22}$$
which is impossible.

Let us mention finally that the indices may be used to determine the exponent to which a number a belongs (mod m). This exponent x is the smallest positive solution of
$$a^x \equiv 1 \pmod{m}$$

Consequently
$$x \operatorname{Ind} a \equiv 0 \pmod{\varphi(m)}$$

and if $\varphi(m)$ and Ind a have the greatest common factor d, one must have
$$x = \frac{\varphi(m)}{d}$$

For instance, the number 3 has the index 16 and since $(16, 22) = 2$ it belongs to the exponent 11 (mod 23).

The index theory is valid only for moduls with primitive roots. This difficulty may be evaded by various methods, by generalization of the theory or by relying on the fact that a congruence can be reduced to prime-power moduls p^α and that for such moduls there exist primitive roots when $p > 2$. A more serious defect, in comparison with logarithms, is that the tables of indices must be computed separately for each modul. Gauss, in a supplement to the *Disquisitiones*, gives a table of indices for the possible moduls up to 100. A monumental achievement is the *Canon arithmeticus* of K. G. J. Jacobi, which contains dual sets of index tables for all prime powers up to 1,000. By means of these tables, a large number of congruence problems for moderate-sized moduls may be solved with great ease.

Problems.

1. Solve the following problems by means of indices:

(a) $7x \equiv 13 \pmod{23}$

(b) $4x \equiv 19 \pmod{23}$

(c) $3x^7 \equiv 11 \pmod{23}$

(d) $11x^3 \equiv 2 \pmod{23}$

(e) $5 \cdot 7^x \equiv 2 \pmod{23}$

(f) $7 \cdot 11^x \equiv 15 \pmod{23}$

(g) To which exponents do 11, 13, and 15 belong (mod 23)?

2. Construct a table of indices (mod 19) and use it to solve the following problems:

(a) $3x \equiv 17 \pmod{19}$

(b) $3x^4 \equiv 4 \pmod{19}$

(c) $13x^7 \equiv 2 \pmod{19}$

(d) $3 \cdot 5^x \equiv 1 \pmod{19}$

(e) To which exponents do the numbers 2, 3, 4, and 5 belong (mod 19)?

12–8. Number theory and the splicing of telephone cables. We have established previously that for each modul m there exist numbers r that belong to the greatest possible exponent $\lambda(m)$, the universal exponent or indicator (mod m). This may be stated in the form that there exist numbers r relatively prime to m such that the remainders (mod m) of the powers

$$r, \quad r^2, \quad r^3, \quad \ldots \tag{12–49}$$

avoid the value $+1$ as long as possible.

We shall now discuss a problem of a very similar character. This time we shall try to determine a number r such that the remainders (mod m) of the series (12–49) avoid both values ±1 as long as possible. For some number r, relatively prime to m, let us say that t is its ±1-*exponent* when it is the smallest possible exponent for which either one of the congruences

$$r^t \equiv \pm1 \pmod{m} \tag{12–50}$$

is fulfilled. Our problem is then to determine the largest possible ±1-exponent $\lambda_0(m)$ that may occur (mod m). We notice that the ±1-exponent of a number can at most be equal to the exponent to which the number belongs (mod m). The general result is:

THEOREM 12–19. The maximal ±1-exponent $\lambda_0(m)$ for a number $m > 2$ has the value

$$\lambda_0(m) = \tfrac{1}{2}\lambda(m) \tag{12–51}$$

when there is a primitive root (mod m), and otherwise

$$\lambda_0(m) = \lambda(m) \tag{12–52}$$

Let us suppose first that there exists a primitive root (mod m). We noticed in Sec. 12–6 that a primitive root can only exist when the congruence

$$x^2 \equiv 1 \ (\text{mod } m)$$

has no other roots than

$$x \equiv \pm 1 \ (\text{mod } m)$$

For a modul with a primitive root one has $\lambda(m) = \varphi(m)$ and we recall that $\varphi(m)$ is even when $m > 2$. From Euler's theorem follows

$$r^{\varphi(m)} = (r^{\frac{1}{2}\varphi(m)})^2 \equiv 1 \ (\text{mod } m)$$

for any r relatively prime to m, consequently

$$r^{\frac{1}{2}\varphi(m)} \equiv \pm 1 \ (\text{mod } m)$$

The maximal ± 1 exponent $\lambda_0(m)$ can therefore at most be equal to $\frac{1}{2}\varphi(m)$. On the other hand it cannot be less than this number, because if t is the ± 1 exponent of a primitive root r, it follows from (12–50) that

$$r^{2t} \equiv 1 \ (\text{mod } m) \tag{12–53}$$

hence $2t$ is divisible by $\varphi(m)$ and t is divisible by $\frac{1}{2}\varphi(m)$. Thus when there exists a primitive root, the maximal ± 1 exponent $\lambda_0(m)$ is determined by (12–51).

To complete the proof of theorem 12–19 we must deduce that $\lambda_0(m) = \lambda(m)$ when there is no primitive root.

Let us take some number r belonging to the exponent $\lambda(m)$. If the ± 1-exponent of r should be $t < \lambda(m)$, the congruence (12–53) would hold and $2t$ would be divisible by $\lambda(m)$. This is only possible when

$$t = \frac{1}{2}\lambda(m)$$

and then one must have

$$r^{\frac{1}{2}\lambda(m)} \equiv -1 \ (\text{mod } m) \tag{12–54}$$

Therefore, if we can find a number r belonging to the exponent $\lambda(m)$ such that the congruence (12–54) does not hold, we will have proved the equality (12–52).

The reasoning must again, as in so many number-theory investigations of this kind, be separated into several cases.

The prime factorization of m may be

$$m = 2^{\alpha_0} p_1^{\alpha_1} p_2^{\alpha_2} \cdots \qquad (12\text{--}55)$$

and we suppose first that there are at least two odd primes p_1 and p_2. Among the φ-functions of the various odd prime powers in m, let $\varphi(p_1^{\alpha_1})$ be one that is divisible by the smallest power of 2 and let us select a number ρ_1 that belongs to the exponent $\frac{1}{2}\varphi(p_1^{\alpha_1})$ (mod $p_1^{\alpha_1}$). The square $\rho_1 = r_1^2$ of a primitive root r_1 can be used. A number r can be defined by the Chinese remainder theorem by the following set of congruences

$$r \equiv 3 \pmod{2^{\alpha_0}}, \qquad r \equiv \rho_1 \pmod{p_1^{\alpha_1}}, \qquad r \equiv r_2 \pmod{p_2^{\alpha_2}}, \cdots$$

where the subsequent numbers $r_2, r_3, \ldots$ are primitive roots for the corresponding prime powers. The exponent to which r belongs (mod m) is equal to the least common multiple

$$[\lambda(2^{\alpha_0}), \ \tfrac{1}{2}\varphi(p_1^{\alpha_1}), \ \varphi(p_2^{\alpha_2}), \ \varphi(p_3^{\alpha_3}), \ \ldots]$$

The manner in which p_1 was chosen insures that this number is equal to $\lambda(m)$. But by the definition of r one has also

$$r^{\frac{1}{2}\varphi(p_1^{\alpha_1})} \equiv \rho_1^{\frac{1}{2}\varphi(p_1^{\alpha_1})} \equiv 1 \pmod{p_1^{\alpha_1}}$$

and since $\frac{1}{2}\varphi(p_1^{\alpha_1})$ divides $\frac{1}{2}\lambda(m)$ this implies

$$r^{\frac{1}{2}\lambda(m)} \equiv 1 \pmod{p_1^{\alpha_1}}$$

which is incompatible with the congruence (12–54).

In the next cases, the number m has the form

$$m = 2^{\alpha_0} \cdot p_1^{\alpha_1} \qquad (12\text{--}56)$$

with $\alpha_1 \geqq 1$ and here $\alpha_0 \geqq 2$ since there shall be no primitive root

For the special cases

$$m = 4 \cdot p_1^{\alpha_1}, \quad m = 8 \cdot p_1^{\alpha_1}$$

one obtains

$$\lambda(m) = [2, \ \varphi(p_1^{\alpha_1})] = \varphi(p_1^{\alpha_1})$$

This shows that the number r defined by the congruences

$$r \equiv 1 \pmod 4, \qquad r \equiv r_1 \pmod{p_1^{\alpha_1}}$$

where r_1 is a primitive root (mod $p_1^{\alpha_1}$), must belong to the exponent $\lambda(m)$ (mod m). Furthermore, the congruence (12–54) cannot hold since

$$r^{\frac{1}{2}\lambda(m)} \equiv 1 \pmod 4$$

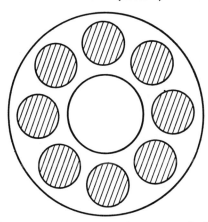

Fig. 12-1. Cross section of one layer of cable.

When we suppose that $\alpha_0 \geqq 4$ in (12–56), we define our r by the congruences

$$r \equiv 3 \pmod{2^{\alpha_0}}, \qquad r \equiv r_1 \pmod{p_1^{\alpha_1}}$$

and as in Sec. 12–5, it follows that r belongs to the exponent $\lambda(m)$ (mod m). In this case the number $\lambda(2^{\alpha_0})$, and therefore $\lambda(m)$, is divisible at least by the second power of 2; from

$$3^2 \equiv 1 \pmod 8$$

one concludes that

$$r^{\frac{1}{2}\lambda(m)} \equiv 3^{\frac{1}{2}\lambda(m)} \equiv 1 \pmod 8$$

which again shows that the congruence (12–54) is not fulfilled.

The same argument with $r = 3$ is applicable in the final case where m is a power of 2.

H. P. Lawther, Jr., in an article in the *American Mathematical Monthly*, has pointed out how such a theory may be applied to introduce a systematic method for splicing telephone cables. The cables for long-distance telephone service are manufactured in concentric layers of insulated wires or conductors. The cable is produced in sections of approximately uniform length, each perhaps 1,000 feet long, and the line is made up of a succession of such sections spliced end to end. At the splices the order of the wires should be mixed up considerably to minimize interference and cross talk, and it is particularly desirable to avoid having two

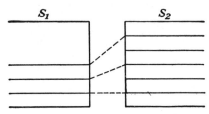

FIG. 12-2.   Splicing arrangement.

conductors that are adjacent in one section adjacent in sections following closely afterward.

For practical purposes the splicing scheme cannot be complicated, and it would seem that the following three rules would embody the utmost in simplicity:

1. The same splicing directions should be used at each intersection.
2. The wires in one concentric layer are spliced to those in the corresponding layer in the next section.
3. When some wire in one section S_1 is spliced to a wire in the next section S_2, the adjacent wire in S_1 should be spliced to the wire in S_2 that is obtained from the one last spliced by counting forward a fixed number s. (See Fig. 12-2 for the case $s = 2$.)

The number s may be called the *spread* of the splicing rule. For instance, if one takes a layer in a cable in which there are 11 wires, the following table gives the splicings for the spreads $s = 2$ and $s = 3$.

$s = 2$	$s = 3$
$1 \to 1$	$1 \to 1$
$2 \to 3$	$2 \to 4$
$3 \to 5$	$3 \to 7$
$4 \to 7$	$4 \to 10$
$5 \to 9$	$5 \to 2$
$6 \to 11$	$6 \to 5$
$7 \to 2$	$7 \to 8$
$8 \to 4$	$8 \to 11$
$9 \to 6$	$9 \to 3$
$10 \to 8$	$10 \to 6$
$11 \to 10$	$11 \to 9$

The numbers in each of the right-hand columns are obtained from the first 1 by adding 2 or 3 each time. Since the arrangement of the wires is circular, the numbers are reduced (mod 11). One sees that the first wire in S_1 always corresponds to the first wire in S_2; this is only an expression for an agreement to let the count in the second section start from the first wire spliced.

In general the splicing table for a spread s will run

$$1 \to 1$$
$$2 \to 1 + s$$
$$3 \to 1 + 2s \qquad (\text{mod } m) \qquad (12\text{-}57)$$
$$\cdot \quad \cdot \quad \cdot \quad \cdot \quad \cdot \quad \cdot$$
$$i \to 1 + (i - 1)s$$
$$\cdot \quad \cdot \quad \cdot \quad \cdot \quad \cdot \quad \cdot$$

where the numbers on the right shall be reduced to their smallest positive remainders (mod m), when m is the number of wires in a cylindrical layer.

It should be observed that the correspondences (12-57) impose a condition limiting the number of acceptable spreads s. As a consequence of the rather trivial fact that two wires in S_1 always are spliced to two distinct wires in S_2, one concludes that all remainders of the numbers on the right in (12–57) must be different. In other words, a congruence

$$1 + is \equiv 1 + js \pmod{m}$$

or

$$(i - j)s \equiv 0 \pmod{m} \tag{12–58}$$

will not be possible except when $i = j$. This is the case only when *the spread s is relatively prime to m;* because if s had a common factor d with m, the congruence (12–58) would be fulfilled whenever i and j differed by m/d or a multiple of this number.

Now let us repeat the splicing process. By the first operation the ith wire in S_1 was connected with the wire number

$$i_2 = 1 + (i - 1)s$$

in S_2. By the second splice this wire leads to wire number

$$i_3 = 1 + (i_2 - 1)s = 1 + (i - 1)s^2$$

and when this is continued the ith wire in the first section, after n splices, will be connected with the wire numbered

$$i_{n+1} \equiv 1 + (i - 1)s^n \pmod{m} \tag{12–59}$$

in the $(n + 1)$ section.

The object of our splicing arrangement was to produce a scatter in the distribution of the connections such that two wires that were adjacent in some section would stay separated as long as possible in the subsequent sections. Let us examine how our scheme behaves in this respect. Two adjacent wires in some section may be numbered i and $i + 1$. According to (12–59), after n splices, they become connected with the wires numbered respectively

$$1 + (i - 1)s^n, \quad 1 + is^n \pmod{m}$$

These two wires are adjacent only if their difference is congruent to $\pm 1 \pmod{m}$, that is, when

$$1 + is^n - (1 + (i - 1)s^n) \equiv \pm 1 \pmod{m}$$

or

$$s^n \equiv \pm 1 \pmod{m}$$

This basic condition leads us directly back to our previous number-theory investigations. It shows that in order to keep onetime adjacent wires separated as far as possible, one must select the spread s as a number whose ± 1-exponent (mod m) has the maximal value $\lambda_0(m)$. In the preceding we have made all necessary preparations for this problem. We know how such a spread s may be determined and also how to compute the value of the corresponding maximal ± 1-exponent $\lambda_0(m)$.

In the paper by Lawther, one finds a table giving the value of $\lambda_0(m)$ for all m's up to $m = 139$, as well as a suitable spread s. For the first small values the table is as follows:

m	λ_0	s	m	λ_0	s
5	2	2	13	6	2
6	1	5	14	3	3
7	3	3	15	4	2
8	2	3	16	4	3
9	3	2	17	8	3
10	2	3	18	3	5
11	5	2	19	9	2
12	2	5	20	4	3

The reader may check some of these results and compute the values of λ_0 for some of the higher moduls m.

Let us make the final observation that when the modul is a prime p, theorem 12–19 shows that

$$\lambda_0 = \frac{p - 1}{2}$$

is the number of sections for which no two wires are adjacent more than once. In comparison with the number m of wires, this is the best possible result obtainable by any method. To see this, one need only notice that when one starts with some wire it can only remain separated from adjacent ones as long as it is possible to find a new pair of wires between which it can be placed at each splice, and altogether there are only $\frac{1}{2}(m - 1)$ such different pairs.

Let us only mention, finally, that there are other ways in which this problem may be handled.

Bibliography

CUNNINGHAM, A. J. C., H. J. WOODALL, and T. G. CREAK: *Haupt-exponents, Residue-indices, Primitive Roots and Standard Congruences*, London, 1922.

JACOBI, K. G. J.: *Canon arithmeticus, sine tabulae quibus exhibentur pro singulis numeris primis vel primorum potestatibus infra* 1000 *numeri ad datos indices et indices ad datos numeros pertinentes*, Berlin, 1839.

KRAITCHIK, M.: *Recherches sur la théorie des nombres*, Paris, 1924.

LAWTHER, JR., H. P.: "An Application of Number Theory to the Splicing of Telephone Cables," *American Mathematical Monthly*, Vol. 42, 81–91 (1935).

CHAPTER 13

THEORY OF DECIMAL EXPANSIONS

13–1. Decimal fractions. In the opening chapter we discussed the number systems and number representations. When the basic group was b, we could write every integer uniquely by means of the powers of b in the form

$$N = (c_n, c_{n-1}, \cdots, c_1, c_0)_b = c_n \cdot b^n \\ + c_{n-1} \cdot b^{n-1} + \cdots + c_1 \cdot b + c_0$$

where the c_i's can take values from 0 to $b - 1$. From a mathematical point of view, it may appear simple to extend this process and use negative powers of b to represent the fractions, as we do with our decimal fractions. Historically, however, this does not seem to have been an easy step. In dealing with the fractions those that are encountered first and the ones that are used most commonly are the very simplest

$$\tfrac{1}{2}, \quad \tfrac{1}{3}, \quad \tfrac{2}{3}, \quad \tfrac{1}{4}, \quad \tfrac{3}{4}, \quad \cdots$$

and it may not appear natural even now to put them in the strait jacket of decimal notation. Furthermore, such representation may even involve conceptual difficulties; for instance, $\tfrac{1}{3}$ is a much more easily understood and explicable concept than the infinite series

$$0.333 \cdots = \frac{3}{10} + \frac{3}{10^2} + \frac{3}{10^3} + \cdots$$

These remarks may explain, at least in part, why the decimal fractions made their appearance relatively late in the history of the number concept and why a similar construction was not achieved in other number systems. As an exception, one should mention the Babylonian sexagesimal system. As we have already stated, it was based on a positional principle but without an absolute determination of values. Thus the symbol ∏ may mean 2 or 2×60 or 2×60^2 and so on, but it can also signify

$$2 \times \frac{1}{60}, \quad 2 \times \frac{1}{60^2}, \cdots$$

Although there was no indication of the integral part of a number or even the power a certain symbol may denote, the system included almost all the practical advantages of the decimal notation. For instance, the common tables of inverses could be considered to give the expansions according to the negative powers of 60. The Babylonian preeminence in extensive and accurate computations must have been induced by this technical perfection of their number system. Probably this is one of the contributory reasons for the persistent use of sexagesimal notations in many later Greek and Arab mathematical texts, particularly in the field of astronomy. We may recall that even Leonardo Pisano, the ardent protagonist of the Hindu positional numbers, when confronted with the necessity of computing a root of a cubic equation with great accuracy in one of his tournament problems, resorts to sexagesimal computations and gives the answer in this notation as

$$x = 1^0 \ 22^{\mathrm{I}} \ 7^{\mathrm{II}} \ 43^{\mathrm{III}} \ 33^{\mathrm{IV}} \ 4^{\mathrm{V}} \ 40^{\mathrm{VI}}$$

There are initial steps towards decimal notation and computations to be found among Arabic and European mathematicians by the fifteenth century. These attempts usually appeared through a desire to eliminate fractional computations by the simplest possible operation, namely, the multiplication of the numbers by powers of 10. For instance, to compute $\sqrt{2}$ one would prefer to take the square root of the number

$$1{,}000\sqrt{2} = \sqrt{2{,}000{,}000}$$

and proceed to calculate with integers.

The first work in which one finds computations with decimal fractions is a collection of reckoning examples by Christoff Rudolff, which appeared in Augsburg in 1530. He uses a decimal notation similar to the modern one, with a bar to separate the integral and fractional parts. In spite of the fact that Rudolff's books, *The Rules of Coss* and *Reckoning Manual,* enjoyed great contemporary popularity, little is known about his life except that he spent most of his time in Vienna as an "amateur of the liberal arts," as he characterizes himself in the foreword to one of his books.

The first systematic presentation of the rules of operations on decimal fractions was given in 1585 in a brief treatise in Dutch, *De Thiende,* written by Simon Stevin. A French translation under the title *La Disme* followed shortly afterward. Simon Stevin (1548–1620) was a curious, many-sided genius who combined his strong scholarly leanings with great ability for practical constructive and organizational tasks. He made inventions of various kinds, became quartermaster general of the Dutch army during the critical period of the Spanish wars, and also was put in charge of the system of dikes and waterworks. He reorganized the system of governmental bookkeeping, computed interest tables, and proposed reforms in weights and measures suggestive of the metric system. His writings on the art of fortifications are well known, but at the same time he translated Diophantos into French and published books on arithmetic, geometry, mechanics, and hydrostatics. He seems to have been the first to formulate explicitly the law of the triangle or parallelogram of forces in mechanics.

La Disme has the subtitle "Teaching how all computations that are met in business may be performed by integers alone without the aid of fractions." The pamphlet opens as follows:

To astrologers, surveyors, measurers of tapestry, gaugers, stereometers in general, mintmasters and to all merchants Simon Stevin sends greeting:

A person who contrasts the small size of this book with your greatness, my most honorable sirs to whom it is dedicated, will think my idea absurd, especially if he imagines that the size of this volume bears the same ratio to human ignorance that its usefulness has to men of your outstanding ability; but in so doing he will have compared the extreme terms of the proportion which may not be done. Let him rather compare the third term with the fourth.

What is it that is here propounded? Some wonderful invention? Hardly that, but a thing so simple that it scarce deserves the name invention; for it is as if some stupid country lout chanced upon great treasure without using any skill in the finding. If any one thinks that, in expounding the usefulness of decimal numbers, I am boasting of my cleverness in devising them, he shows without doubt that he has neither the judgment nor the intelligence to distinguish simple things from difficult, or else that he is jealous of a thing that is for the common good. However this may be, I shall not fail to mention the usefulness of these numbers, even in the face of this man's empty calumny. But, just as the mariner who has found by chance an unknown isle, may declare all its riches to the king, as, for instance, its having beautiful fruits, pleasant plains, precious minerals, etc., without its being imputed to him as conceit; so may I speak freely of the great usefulness of this invention, a usefulness greater than I think any of you anticipates, without constantly priding myself on my achievements.[1]

The introduction of the decimals was an innovation for which the time was ripe, and after the publication of *La Disme* one sees a rapid increase in their use. In numerical tables the advantages of decimals over ordinary fractions are particularly evident. Shortly before the appearance of *La Disme*, Stevin had computed and published a set of interest tables, and one may well conjecture that this had led him to an appreciation of the simplicity of decimal procedures. The invention of logarithms and the introduction of logarithmic tables shortly afterwards undoubtedly contributed greatly to the progress of decimals.

Stevin presaged another practical innovation in a statement toward the end of the pamphlet

In view of the great usefulness of the decimal division, it would be a praiseworthy thing if the people would urge having this put into effect so that in addition to the common divisions of the measures, weights and money, that now exist, the state would declare the decimal division of the

[1] Translation by V. Sanford, in D. E. Smith, *A Source Book in Mathematics.*

large units legitimate to the end that he who wished might use them. It would further this cause also, if all new money should be based on this system of primes, seconds, thirds, etc. [tenths, hundreds, thousands]. If this is not put into operation as soon as we might wish, we have the consolation that it will be of use to posterity, for it is certain that if men of the future are like men of the past, they will not always be neglectful of a thing of such great value.

Not all of Stevin's reform proposals had the same ultimate success. In one of his works on geography he makes strong propaganda for Dutch as a world language, and he demonstrates its supremacy over other languages in the briefness of expression and in its superabundance of words of one syllable.

13–2. The properties of decimal fractions. In computing with decimals it soon became apparent that the numbers fall into several distinct categories with respect to their expansions. For some the expansion might break off after a finite number of terms; for others it might be infinite. In the latter case the digits could progress without any discernible law, or from a certain point on they might repeat themselves periodically, forming *circulating decimals*. Certain rules for the expansion of rational numbers into decimals had already been propounded by earlier mathematicians, but not until Gauss were those indispensable tools of number theory created that were required for a systematic exploration. In the sixth and next to last section of the *Disquisitiones*, Gauss considers the use of his results in certain applications, among them the properties of decimal expansions of fractions.

The characterization of the numbers with a finite decimal expansion is quite simple; the main idea has in reality been presented already in our discussion of the Babylonian mathematical tablets. As a general remark for the subsequent considerations, let us observe that when dealing with the properties of decimal expansions, we shall usually disregard the integral part of a number since it has no influence on the decimals.

When some number has a finite expansion

$$r = 0, a_1, a_2, \cdots, a_n = \frac{a_1}{10} + \frac{a_2}{10^2} + \cdots + \frac{a_n}{10^n}$$

it may be written

$$r = \frac{a_1 \cdot 10^{n-1} + a_2 \cdot 10^{n-2} + \cdots + a_n}{10^n} = \frac{p}{q}$$

Therefore r is a rational number whose denominator q in the reduced form contains only prime factors of 10, that is, 2 and 5. For instance,

$$0.1375 = \frac{1,375}{10,000} = \frac{11}{80} = \frac{11}{2^4 \cdot 5}$$

Conversely let us take a fraction

$$r = \frac{p}{2^\alpha \cdot 5^\beta}$$

If, for example, $\alpha \geqq \beta$, we may multiply both numerator and denominator by $5^{\alpha-\beta}$ so that

$$r = \frac{5^{\alpha-\beta}p}{10^\alpha}$$

and this shows that r can be written with a finite decimal expansion.

THEOREM 13–1.   The numbers with a finite decimal expansion are the fractions

$$r = \frac{p}{2^\alpha \cdot 5^\beta} \tag{13-1}$$

It is not difficult to see that when r is in reduced form (13–1), the number of decimals in the expansion is the larger of the two exponents α and β.

In general, let us say that a number is *regular* with respect to some base number b when it can be expanded in the corresponding number system with a finite number of negative powers of b. Through exactly the same argument as before, one concludes that the regular numbers are the fractions

$$r = \frac{p}{q}$$

where q contains no other prime factors than those that divide b. Then there exists some exponent n such that q divides b^n, and when r is reduced and n is the smallest exponent that can be used for this purpose, the number of "decimal" terms is exactly n.

Since we are mainly interested in the number-theory properties that regulate the expansions into decimal fractions, we shall suppose that the reader is familiar with the essential principles of decimals. In particular, we know that every real number can be expressed by a decimal fraction, and this expansion can be performed in only one way. The single exception to this last statement occurs for numbers with a finite expansion. Here one can diminish the last digit by one unit and continue with an infinite series of nines, as for instance, in the examples

$$1.00 \cdots = 0.999 \cdots, \qquad 0.375 = 0.374999 \cdots$$

Since we have already completed the discussion of numbers with finite expansions this anomaly need not concern us here.

We shall examine the decimal expansions one obtains for the various reduced fractions

$$r = \frac{m}{n}, \qquad (m, n) = 1 \quad (13\text{–}2)$$

where the denominator n is some fixed integer. Gauss calls the decimal sequence of a number, disregarding the integral part, the *mantissa* of the number. This is a term that was first introduced by Briggs in connection with his logarithms to the base 10, and he uses it in the sense of a "minor part" or "appendix." As we have already stated, the mantissa of a number is the main object of our subsequent studies.

Two fractions m/n and l/n in (13–2) have the same mantissa when their difference

$$\frac{m}{n} - \frac{l}{n} = k$$

is an integer, in other words when

$$m \equiv l \pmod{n}$$

Therefore, to obtain the various mantissas for the fractions (13-2), we can limit ourselves to those fractions in which the numerator m is one of the $\varphi(n)$ remainders less than and relatively prime to n. For instance, for the denominator $n = 18$, the fractions

$$\tfrac{1}{18}, \quad \tfrac{5}{18}, \quad \tfrac{7}{18}, \quad \tfrac{11}{18}, \quad \tfrac{13}{18}, \quad \tfrac{17}{18}$$

will exhaust all possibilities for the mantissas.

Another simple but essential remark is the following: When the mantissa of the fraction m/n is known, one finds the mantissa of the fraction $10 \cdot m/n$ by dropping the first digit on the left. For instance

$$\tfrac{1}{7} = 0.142857 \cdots$$

has the mantissa $142857 \ldots$, while the mantissa of $\tfrac{10}{7}$ is $428571 \ldots$.

On the basis of these observations, it is not difficult to prove that the decimal expansion of a rational number m/n is always periodic, *i.e.*, that the mantissa after a certain number of terms consists of groups of digits that keep repeating themselves indefinitely. For instance,

$$\tfrac{57}{165} = 0.3\overline{45}4545 \cdots$$

We consider the series of fractions

$$\frac{m}{n}, \quad \frac{10m}{n}, \quad \frac{10^2m}{n}, \ldots \tag{13-3}$$

Here the mantissa of the first fraction will produce those of all the subsequent ones by leaving out successively one, two, and so on, digits from the left. But in the unlimited sequence of numerators of the fractions (13-3)

$$m, \quad 10m, \quad 10^2m, \ldots$$

the numbers cannot all be incongruent (mod n). Consequently, there exists some first exponent s such that

$$m \cdot 10^s \equiv m \cdot 10^{s+t} \pmod{n} \tag{13-4}$$

and the two fractions

$$\frac{m \cdot 10^s}{n}, \qquad \frac{m \cdot 10^{s+t}}{n}$$

have the same mantissa. This means that leaving out s digits and $s + t$ digits from the mantissa of m/n will produce the same sequence. We conclude, therefore, that in the decimal expansion of m/n the digits will repeat themselves periodically after s terms, in groups of length t.

Since in (13–2) we made the assumption that m and n were relatively prime, the factor m in (13–4) can be canceled to give the equivalent congruence

$$10^s \equiv 10^{s+t} \pmod{n}$$

Conversely, let us take some periodic decimal fraction

$$r = 0.a_1a_2 \cdots a_s \quad b_1b_2 \cdots b_t \quad b_1b_2 \cdots b_t \cdots \qquad (13–5)$$

where the period has the length t and begins after the s first terms. When (13–5) is multiplied by 10^s and 10^{s+t} one obtains

$$10^{s+t}r = a_1a_2 \cdots a_sb_1b_2 \cdots b_t.b_1b_2 \cdots b_t \cdots$$

$$10^sr = a_1a_2 \cdots a_s.b_1b_2 \cdots b_t \cdots$$

The two numbers on the right have the same mantissa so that their difference A is an integer. By subtracting one from the other, we find

$$(10^{s+t} - 10^s)r = A$$

and we conclude that

$$r = \frac{A}{10^{s+t} - 10^s} = \frac{m}{n}$$

is rational. Furthermore, it is clear that 10^s and 10^{s+t} are the smallest powers of 10 such that $r \cdot 10^s$ and $r \cdot 10^{s+t}$ have the same mantissa. We conclude therefore:

THEOREM 13–2. Any periodic decimal fraction represents a rational number, and conversely any rational number has a periodic decimal expansion.   When a rational number

$$\frac{m}{n}, \qquad (m, n) = 1$$

is expanded, the period begins after s terms and has a length t, where s and t are the smallest numbers such that

$$10^s \equiv 10^{s+t} \pmod{n} \qquad (13\text{--}6)$$

This condition shows the further interesting fact that the length of the period as well as the point at which it starts depends only on the denominator n and not on the numerator m.

Examples.

1. Let us consider the expansions with denominator 18.   For the powers of 10 (mod 18), one finds simply

$$1, \quad 10, \quad 10^2 \equiv 10$$

so that $s = 1$ and $t = 1$.   This is confirmed by the decimal fractions

$$\tfrac{1}{18} = .055 \cdots, \qquad \tfrac{11}{18} = .611 \cdots$$
$$\tfrac{5}{18} = .277 \cdots, \qquad \tfrac{13}{18} = .722 \cdots$$
$$\tfrac{7}{18} = .388 \cdots, \qquad \tfrac{17}{18} = .944 \cdots$$

2. In a second illustration we take the fractions with denominator 84.   The remainders of the powers of 10 (mod 84) are found to be

$$1, \quad 10, \quad 16, \quad -8, \quad 4, \quad 40, \quad -20, \quad -32, \quad 16$$

The period therefore begins after two terms and has the length 6; for instance,

$$\tfrac{37}{84} = .44 \ \overline{047619} \ 047619 \cdots$$

When the denominator n of the fraction to be expanded has no factors 2 or 5, the conditions become simpler.   Since n is relatively prime to 10, we can cancel in the congruence (13–6) to obtain

$$10^t \equiv 1 \pmod{n} \qquad (13\text{--}7)$$

This shows that we have $s = 0$ and the period starts with the first decimal, or as one sometimes says, the expansion is *purely periodic*.   The length of the period is the exponent to which 10 belongs (mod n). Conversely, it is clear that a congruence (13-7) can hold only when n and 10 are relatively prime so that we may say:

THEOREM 13-3. The decimal expansion of an irreducible fraction m/n is purely periodic if and only if n has no prime factors 2 and 5, and in this case the length of the period is equal to the divisor t of $\varphi(n)$ to which 10 belongs (mod n).

Example.

Let us take $n = 7$. The remainders of the successive powers of 10 (mod 7) are

$$1, \quad 3, \quad 2, \quad 6, \quad 4, \quad 5, \quad 1$$

This shows that 10 belongs to the exponent 6 (mod 7), or, in other words, 10 is a primitive root (mod 7). All fractions with denominator 7 have periods of length 6; for instance,

$$\tfrac{2}{7} = .\overline{285714} \cdots$$

We return for a moment to the general case where the denominator n in the fraction may have factors 2 and 5, and write

$$n = n_0 \cdot 2^\alpha \cdot 5^\beta, \qquad (n_0, \ 10) = 1$$

The preceding result may be used to reformulate the criteria in theorem 13-2. When one multiplies the fraction m/n by 10^μ, where μ is the larger of the exponents α and β, the resulting fraction

$$10^\mu \, \frac{m}{n} = \frac{2^{\mu-\alpha} \cdot 5^{\mu-\beta} \cdot m}{n_0}$$

has a denominator relatively prime to 10 and μ is the lowest exponent by means of which this can be achieved. Theorem 13-3 gives us then:

THEOREM 13-4. When the denominator of a fraction m/n has the form

$$n = n_0 \cdot 2^\alpha \cdot 5^\beta, \qquad (n_0, \ 10) = 1$$

the period in the decimal expansion of m/n begins after μ terms, where μ is the larger of α and β, and the length of the period is the exponent to which 10 belongs (mod n_0).

Examples.

Let us reconsider our previous examples from this new point of view. When $n = 18 = 2 \cdot 9$, the period must begin after the first term and have the length 1 since

$$10 \equiv 1 \pmod 9$$

When $n = 84 = 2^2 \cdot 21$, the period starts after the second decimal and has the length 3 since one finds that 10 belongs to the exponent 6 (mod 21).

We have mentioned that in the expansion of an irreducible fraction m/n in decimals all periods begin at the same point and have the same length for a given n. Let us discuss briefly the interrelation between the various periods defined by n. We may multiply m/n by a suitable power of 10 so that the expansion becomes purely periodic, or, equivalently, n shall be assumed relatively prime to 10.

The general situation can best be explained on the basis of some examples.

Examples.

1. We first take the decimal expansions of fractions with denominator 7. Since 10 belongs to the exponent 6 (mod 7), the period is 6 and

$$\tfrac{1}{7} = .\overline{142857} \cdots$$

This fraction is multiplied by the successive powers of 10 and the integral parts discarded. There results a set of decimal fractions whose mantissas are derived from that of $\frac{1}{7}$ by leaving out one, two, and so on, digits on the left. Since the powers of 10 (mod 7) have the remainders

$$1, \quad 3, \quad 2, \quad 6, \quad 4, \quad 5, \quad 1$$

we conclude that

$$\tfrac{1}{7} = .\overline{142857} \cdots, \qquad \tfrac{6}{7} = .\overline{857142} \cdots$$

$$\tfrac{3}{7} = .\overline{428571} \cdots, \qquad \tfrac{4}{7} = .\overline{571428} \cdots$$

$$\tfrac{2}{7} = .\overline{285714} \cdots, \qquad \tfrac{5}{7} = .\overline{714285} \cdots$$

All possible mantissas for the denominator 7 can therefore be obtained from one of them by permuting the digits in the period cyclically. It is evident that the same situation will prevail whenever 10 is a primitive root (mod n). An index table in which the numbers are arranged according to their indices will give the information as to which period appears for a prescribed fraction

2. Next let us take an example where 10 is not a primitive root (mod n), for instance, $n = 13$. In this case the remainders of the powers of 10 (mod 13) are

$$1, \quad 10, \quad 9, \quad 12, \quad 3, \quad 4, \quad 1$$

and from the expansion of $\frac{1}{13}$ one consequently finds

$$\frac{1}{13} = .\overline{076923} \cdots, \qquad \frac{12}{13} = .\overline{923076} \cdots$$

$$\frac{10}{13} = .\overline{769230} \cdots, \qquad \frac{3}{13} = .\overline{230769} \cdots$$

$$\frac{9}{13} = .\overline{692307} \cdots, \qquad \frac{4}{13} = .\overline{307692} \cdots$$

Here we have only half of the twelve reduced proper fractions with denominator 13. The numerator 2 is not among them so that we multiply 2 by the powers of 10 (mod 13). The remainders are

$$2, \quad 7, \quad 5, \quad 11, \quad 6, \quad 8, \quad 2$$

and, correspondingly, one has the cyclic family of expansions

$$\frac{2}{13} = .\overline{153846} \cdots, \qquad \frac{11}{13} = .\overline{846153} \cdots$$

$$\frac{7}{13} = .\overline{538461} \cdots, \qquad \frac{6}{13} = .\overline{461538} \cdots$$

$$\frac{5}{13} = .\overline{384615} \cdots, \qquad \frac{8}{13} = .\overline{615384} \cdots$$

In the general case the situation is analogous. When the number 10 belongs to the exponent t, the $\varphi(n)$ mantissas of the fractions with the denominator n will fall into $\varphi(n)/t$ families, the periods are of length t and within each family the mantissas are obtained by cyclical permutations as above. For instance, 10 belongs to the exponent 5 (mod 41); hence, the period of the fractions with denominator 41 is equal to 5 and the mantissas fall into 8 cyclical classes.

There exist tables that give the classes of mantissas for all numbers not divisible by 2 or 5 up to certain limits. Gauss in an appendix to the *Disquisitiones* gave such a table, which he later enlarged. The most complete tables of this kind are due to H. Goodwin and include the mantissas for all denominators up to 1,024, but these tables are now so rare that they are practically unavailable.

Very extensive tables have been computed to determine the exponents to which the number 10 belongs for various moduls.

In arranging these tables one makes use of the fact, which is easily proved, that if 10 belongs to the exponent a (mod m) and the exponent b (mod n), then for the least common multiple of m and n as modul, it belongs to an exponent that is the l.c.m. of a and b. It is sufficient, therefore, to tabulate the exponents to which 10 belongs for the various prime powers p^α. For $p^\alpha < 10,000$, such tables have been constructed by A. J. C. Cunningham and coworkers. For prime moduls, still more extensive tables with $p < 120,000$ have been computed; one should mention particularly those by W. Shanks.

One may be interested in determining the denominators that yield short periods. Such a study is facilitated through the factorization of the various numbers $10^k - 1$ into prime factors. For the first few exponents one obtains

$$10 - 1 = 3^2 \qquad\qquad 10^5 - 1 = 3^2 \cdot 41 \cdot 271$$
$$10^2 - 1 = 3^2 \cdot 11 \qquad\qquad 10^6 - 1 = 3^3 \cdot 7 \cdot 11 \cdot 13 \cdot 37$$
$$10^3 - 1 = 3^3 \cdot 37 \qquad\qquad 10^7 - 1 = 3^2 \cdot 239 \cdot 4,649$$
$$10^4 - 1 = 3^2 \cdot 11 \cdot 101 \qquad 10^8 - 1 = 3^2 \cdot 11 \cdot 73 \cdot 101 \cdot 137$$
$$10^9 - 1 = 3^4 \cdot 37 \cdot 333,667$$

From this table one concludes, for instance, that 7 and 13 are the only primes whose periods have the length 6, while 239 and 4,649 have the period 7, and so on.

In the preceding we have limited ourselves exclusively to the case where the base of the number system is 10, but it is evident that the results we have obtained are valid, with very small modifications, for arbitrary base numbers. For example, when examining the Babylonian tables of inverses, one may wish to know which denominators correspond to short periods in the sexagesimal system. The following prime factorizations yield this information:

$$60 - 1 = 59 \qquad\qquad 60^3 - 1 = 7 \cdot 59 \cdot 523$$
$$60^2 - 1 = 59 \cdot 61 \qquad 60^4 - 1 = 13 \cdot 59 \cdot 61 \cdot 277$$
$$60^5 - 1 = 11 \cdot 59 \cdot 1,198,151$$

Problems.

1. Find the length of the decimal period for fractions with the denominators

$$n = 17, \qquad n = 31, \qquad n = 39, \qquad n = 43$$

and find the corresponding families of mantissas.

2. At which point does the period begin when $n = 10!$?

3. Find all numbers whose periods are 6 and 12 in a number system with the base 2.

4. In which number systems does a prime power p^α give a period of length 2?

Bibliography

CUNNINGHAM, A. J. C., H. J. WOODALL, and T. G. CREAK: *Haupt-exponents, Residue-indices, Primitive Roots and Standard Congruences*, London, 1922.

GOODWIN, H.: *A Table of Circles arising from the Division of a Unit or Any Other Whole Number by All the Integers from 1 to 1024, being all the pure decimal quotients that can arise from this source*, London, 1823.

———: *A Tabular Series of Decimal Quotients of All Proper Vulgar Fractions of which, when in their lowest terms, neither the numerator nor the denominator is greater than 1000*, London, 1823.

STEVIN, SIMON: *La Disme.* Translation in D. C. Smith, *A Source Book in Mathematics*, McGraw-Hill Book Company, Inc., New York, 1929.

CHAPTER 14

THE CONVERSE OF FERMAT'S THEOREM

14–1. The converse of Fermat's theorem. Fermat's theorem states that for every number a not divisible by the prime p, the congruence

$$a^{p-1} \equiv 1 \pmod{p}$$

is satisfied. It is natural to investigate conversely whether the fact that some congruence of this kind holds implies that the modul is a prime.

In general such a conclusion is not valid. There exist numbers a and n such that

$$a^{n-1} \equiv 1 \pmod{n}, \qquad a \not\equiv 1 \pmod{n} \tag{14–1}$$

without n being a prime. Several writers have made this observation; for instance, F. Sarrus (1819) noted the congruence

$$2^{340} \equiv 1 \pmod{341}$$

where the modul $341 = 11 \cdot 31$ is composite. Another example is

$$3^{90} \equiv 1 \pmod{91}, \qquad 91 = 7 \cdot 13$$

and numerous other instances may be given.

However, by imposing additional restrictions on the number a in the congruence (14–1), it is possible to express a converse form of the theorem of Fermat. This observation was first made and applied by the French specialist in number theory E. Lucas (1876). His original theorem was

THEOREM 14–1. When for some number a the congruence

$$a^{n-1} \equiv 1 \pmod{n} \tag{14–2}$$

holds, while no similar congruence with a lower exponent

$$a^t \equiv 1 \pmod{n}, \quad n - 1 > t > 0 \tag{14-3}$$

is fulfilled, the modul n is a prime.

On the basis of our previous results the proof is immediate. The condition of the theorem states that the number a belongs to the exponent $n - 1 \pmod{n}$. But the highest exponent to which a can belong is $\varphi(n)$. We recall further that

$$\varphi(n) = n \left(1 - \frac{1}{p_1}\right) \cdots \left(1 - \frac{1}{p_r}\right)$$

where the p_i's are the different primes dividing n so that

$$\varphi(n) \leqq n \left(1 - \frac{1}{p_1}\right) = n - \frac{n}{p_1} \leqq n - 1$$

This shows that one can have

$$\varphi(n) = n - 1$$

only when $n = p_1$ is a prime.

When it comes to the actual verification that a number is a prime, Lucas's theorem is not very practical in the form in which it stands. However through a few further remarks it may be effectively improved upon.

First, if a congruence (14-3) should hold for some exponent t, the number a will belong to an exponent d less than $n - 1$. According to the congruence (14-2), d would divide $n - 1$. Instead of investigating *all* congruences (14-3), therefore, it is sufficient to examine whether such a congruence can hold when the exponent is a proper divisor of $n - 1$.

Second, one need not consider *all* divisors t of $n - 1$, because when the congruence (14-3) holds for some t, it must be fulfilled for all multiples of t. This leads us to reformulate the theorem of Lucas as follows:

THEOREM 14-2. Let n be some integer and

$$q_1, \quad q_2, \quad \cdots, \quad q_s$$

the different prime factors dividing $n - 1$.   If for some number a the congruence

$$a^{n-1} \equiv 1 \pmod{n}$$

holds, while none of the congruences

$$a^{\frac{n-1}{q_i}} \equiv 1 \pmod{n} \qquad i = 1, 2, \cdots, s \qquad (14\text{–}4)$$

are fulfilled, the number n is a prime.

To prove this statement it suffices to observe that any divisor of $n - 1$ must divide one of the *maximal* divisors

$$\frac{n - 1}{q_1}, \quad \cdots, \quad \frac{n - 1}{q_s}$$

so that when a congruence (14–3) holds for some exponent t dividing $n - 1$, at least one of the congruences (14–4) must be satisfied.

To apply the theorem one selects some number a, usually small, for instance, $a = 2$ or $a = 3$, and computes the remainder of the power a^{n-1} for the modul n.   If the congruence (14–2) should not be fulfilled, one concludes that n is not a prime.   The method itself is quite practical.   However, it has the disadvantage that when it has been used to decide that some particular number is composite, one is left in the rather curious position of having no clue to what the factors may be.

In the other alternative, if it should turn out that the congruence (14–2) is true for some number a, the prime factors of $n - 1$ must be found and the congruences (14–4) examined.   If none of them hold, n is a prime; when one of them is fulfilled, the method gives no final decision and one can try the same procedure for some other number a.   When there are few different prime factors of $n - 1$, the number of congruences (14–4) to be investigated is small.   This is the case for many of the larger special numbers to which the method has been applied.

Lucas's converse form of Fermat's theorem involves the com-

putation of the remainders (mod n) of the high powers of a and
it is essential that the work be organized in the most effective
manner. One suitable procedure will be illustrated first on the
very simple example $n = 143$ and $a = 2$.

A	B	C	
142	$2{,}116 \equiv -29$		
71	$2{,}025 \equiv\;\;\; 23$	46	
35	$3{,}481 \equiv\;\;\; 49$	$98 \equiv -45$	(mod 143)
17	$900 \equiv\;\;\; 42$	$84 \equiv -59$	
8	$256 \equiv -30$		
4	16		
2	4		
1	2		

The left-hand column is constructed first. It begins with the
top entry $n - 1 = 142$, and each successive number is the quotient of the preceding when divided by 2. These entries are the
exponents of the various powers to which $a = 2$ shall be raised in
the second column. Here one proceeds from the bottom upward.
The lowest entries are a raised to the powers 1, 2, 4, and 8, respectively. To obtain the 17th power, the preceding entry, -30, is
squared and reduced (mod n) to give 42 as the remainder of the
16th power; this is multiplied by 2 and entered in the third column
as the remainder -59 of the 17th power. Similarly by squaring
-59, one has the remainder 49 of the 34th power, which is doubled
and entered in the third column for the 35th power. Since finally
the computations show that the 142nd power is not congruent to 1,
the conclusion is that the number 143 is composite.

This example, although trivial, gives the key to the general
setup, which is well adapted to machine computation. The
third column is largely superfluous and it has been included above
only for greater clarity of explanation.

We shall give some examples that illustrate the power of the
method.

Examples.

1. We take $n = 700,001$ and use the auxiliary number $a = 3$. The table of computations takes the form:

A	B	A	B
700,000	1	683	564,086
350,000	700,000	341	111,831
175,000	197,937	170	2,222
87,500	182,523	85	686,051
43,750	413,705	42	546,504
21,875	70,917	21	238,260
10,937	481,554	10	59,049
5,468	37,250	5	243
2,734	426,023	2	9
1,367	282,506	1	3

This shows that our number is likely to be a prime. The prime factors of 700,000 are 2, 5, and 7 and the corresponding quotients

$$350,000, \qquad 140,000, \qquad 100,000$$

By similar computations one obtains

$$3^{350,000} \equiv 700,000$$
$$3^{100,000} \equiv 591,336 \qquad (\text{mod } 700,001)$$
$$3^{140,000} \equiv 425,344$$

Since none of these remainders is 1 we conclude that 700,001 is a prime.

2. We want to determine the character of the number 373,831. We use $a = 2$ and obtain the following table of residues:

A	B	A	B
373,830	104,740	365	333,424
186,915	110,115	182	261,278
93,457	327,152	91	323,778
46,728	182,017	45	236,742
23,364	313,895	22	82,163
11,682	140,353	11	2,048
5,841	345,890	5	32
2,920	111,522	2	4
1,460	124,779	1	2
730	205,672		

This shows that the number is composite.

Both of the last two examples are trivial in the sense that the numbers examined are within the limits of the prime tables. The importance of the method lies, of course, in the fact that it is applicable to numbers of arbitrary size.

D. H. Lehmer and P. Poulet have contributed a valuable adjunct to the method of testing primality by means of the converse of Fermat's theorem. Through various ingenious devices they have constructed tables containing the *composite* numbers n up to 100,000,000 for which the congruence

$$2^{n-1} \equiv 1 \pmod{n} \tag{14-5}$$

is fulfilled, and for each such n a prime factor is given. The tables of Poulet are somewhat simpler to use than those of Lehmer, which leave out numbers n with prime factors not exceeding 313.

To determine whether a number n within the limit of the tables is a prime, one checks first by Poulet's tables whether it is one of the exceptional composite numbers for which (14-5) holds. When this is not the case, n is a prime if and only if the congruence (14-5) is satisfied, and through our previous method the test can be performed fairly quickly. (See Supplement.)

Problems.

Check by the converse theorem of Fermat whether the following numbers are composite or prime:

1. $n = 2^{16} + 1$ 3. $n = 300,301$

2. $n = 1,111,111$ 4. $n = 1,234,567$

14–2. Numbers with the Fermat property. We have mentioned in the last section that for certain composite numbers n there may exist numbers a for which

$$a^{n-1} \equiv 1 \pmod{n} \tag{14-6}$$

Much more remarkable is the fact that one can find numbers n that are not prime such that Fermat's congruence (14-6) is satisfied for *every* number a relatively prime to n. Numbers of this kind shall be said to have the *Fermat property* or, for short, we may call them *F numbers*.

The existence of F numbers was first pointed out by R. D. Carmichael (1909). The smallest among them is

$$561 = 3 \cdot 11 \cdot 17 \tag{14-7}$$

and they are on the whole quite rare. Below 2,000 there are only two others, namely,

$$1,105 = 5 \cdot 13 \cdot 17, \qquad 1,729 = 7 \cdot 13 \cdot 19$$

Relatively few investigations on F numbers have been made, and little is known about them beyond the properties we shall deduce in the following.

From the definition of the universal exponent or indicator $\lambda(n)$ of the number n, we conclude first that a congruence (14-6) can hold for all a's relatively prime to n only when the exponent $n - 1$ is divisible by $\lambda(n)$. This leads to the basic criterion for an F number:

THEOREM 14-3. The necessary and sufficient condition for a number n to have the Fermat property is that

$$n \equiv 1 \pmod{\lambda(n)} \tag{14-8}$$

where $\lambda(n)$ is the universal exponent (mod n).

To illustrate the application of this theorem let us verify that the number 561 in (14-7) actually is an F number. We recall the formula for $\lambda(n)$ and find

$$\lambda(561) = [\varphi(3), \quad \varphi(11), \quad \varphi(17)] = [2, \quad 10, \quad 16] = 80$$

and corresponding to (14-8) one has

$$561 \equiv 1 \pmod{80}$$

Certain simple properties of the F numbers flow directly from the criterion in theorem 14-3:

THEOREM 14-4. A number with the Fermat property is odd and equal to a product of different prime factors

$$n = p_1 p_2 \cdots p_s \tag{14-9}$$

where the number of primes is at least three.

These results were given first by Carmichael and for the examples of F numbers given above they are evidently fulfilled.

For the proof of theorem 14–4 we observe that according to the congruence (14–8), the two numbers n and $\lambda(n)$ cannot have any common factor. The number $\lambda(n)$, as we have mentioned several times before, is always even when $n > 2$, consequently n is odd. Next let p^α be the highest power of some prime p that divides n. The indicator $\lambda(n)$ is divisible by

$$\varphi(p^\alpha) = p^{\alpha-1}(p - 1)$$

so that if $\alpha > 1$, both n and $\lambda(n)$ would have the factor p. This establishes that an F number has the form (14–9). It remains to prove that n cannot be the product of two different primes. In that case one would have

$$n = p_1 p_2$$

and the indicator would be

$$\lambda(n) = [p_1 - 1, \quad p_2 - 1]$$

To fulfill the congruence (14–8) the number

$$n - 1 = p_1 p_2 - 1 = (p_1 - 1)p_2 + p_2 - 1$$

must be divisible by $p_1 - 1$. This is possible only when $p_1 - 1$ divides $p_2 - 1$, and in the same manner one concludes that $p_2 - 1$ must divide $p_1 - 1$. Consequently

$$p_1 - 1 = p_2 - 1$$

and the two prime factors would be equal, contrary to our previous conclusion.

On the basis of theorem 14–4 the condition (14–8) for an F number may be reformulated. When n is a product (14–9) of different prime factors, the corresponding indicator is the least common multiple

$$\lambda(n) = [p_1 - 1, \cdots, p_s - 1]$$

The single congruence (14–8) may therefore be replaced by the family of congruences

$$n \equiv 1 \qquad (\text{mod } p_i - 1) \qquad i = 1, 2, \ldots, s \qquad (14\text{–}10)$$

The F numbers, as we stated, are quite scarce. In the tables by Poulet, which were mentioned in the preceding section, the composite numbers n were listed for which the congruence

$$2^{n-1} \equiv 1 \ (\text{mod } n)$$

holds. The F numbers must be found among these entries; they have been especially marked with an asterisk and below 100,000,-000 one counts a total of 250.

A method for constructing F numbers has been given by J. Chernick (1939) and a similar method by S. Sispánov (1941). The three examples we quoted all contain three prime factors, and we shall discuss this case in some detail.

Let

$$n = p_1 p_2 p_3 \qquad (14\text{–}11)$$

be such an F number. Our first observation is contained in the lemma:

Any two of the three numbers

$$p_1 - 1, \quad p_2 - 1, \quad p_3 - 1 \qquad (14\text{–}12)$$

have the same greatest common divisor.

For instance, in the example

$$1{,}729 = 7 \cdot 13 \cdot 19$$

one finds

$$(6, 12) = (6, 18) = (12, 18) = 6$$

To prove the lemma it is sufficient to show that the g.c.d. of any pair of the numbers in (14–12), for instance,

$$d = (p_1 - 1, \quad p_2 - 1)$$

divides the two others, namely,

$$(p_1 - 1, \quad p_3 - 1), \qquad (p_2 - 1, \quad p_3 - 1)$$

or, equivalently, that d divides $p_3 - 1$. But from the definition of d, it follows that

$$p_1 \equiv 1, \qquad p_2 \equiv 1 \qquad (\text{mod } d)$$

so that according to (14–10) and (14–11)

$$n = p_1 p_2 p_3 \equiv p_3 \equiv 1 \ (\text{mod } d)$$

as required.

We use the lemma to write

$$p_1 - 1 = dP_1, \qquad p_2 - 1 = dP_2, \qquad p_3 - 1 = dP_3$$

or

$$p_1 = 1 + dP_1, \qquad p_2 = 1 + dP_2, \qquad p_3 = 1 + dP_3 \quad (14\text{--}13)$$

where the numbers P_1, P_2, P_3 are relatively prime in pairs. This yields

$$\lambda(n) = [p_1 - 1, \ p_2 - 1, \ p_3 - 1] = [dP_1, \ dP_2, \ dP_3] = dP_1 P_2 P_3$$

and the basic condition (14–8) for an F number takes the form

$$p_1 p_2 p_3 \equiv 1 \ (\text{mod } dP_1 P_2 P_3)$$

Here we substitute the values (14–13) and expand the left-hand product to obtain

$$d^3 P_1 P_2 P_3 + d^2(P_1 P_2 + P_1 P_3 + P_2 P_3)$$
$$+ \ d(P_1 + P_2 + P_3) + 1 \equiv 1 \ (\text{mod } dP_1 P_2 P_3)$$

This reduces first to

$$d^2(P_1 P_2 + P_1 P_3 + P_2 P_3) + d(P_1 + P_2 + P_3)$$
$$\equiv 0 \ (\text{mod } dP_1 P_2 P_3)$$

and then again to

$$d(P_1 P_2 + P_1 P_3 + P_2 P_3) \equiv -(P_1 + P_2 + P_3)$$
$$(\text{mod } P_1 P_2 P_3) \quad (14\text{--}14)$$

To construct F numbers from this condition we proceed as follows. Three positive numbers $P_1, P_2,$ and P_3, relatively prime in pairs, are selected. The suitable values for d corresponding to them are solutions of the linear congruence (14–14). Clearly the coefficient of d is relatively prime to the modul so that there exists a unique smallest positive solution d_0, and the corresponding general solution becomes

$$d = d_0 + tP_1P_2P_3 \qquad (14\text{–}15)$$

Since we assume that d is positive, t runs through the series $0, 1, 2, \ldots$. When the value (14–15) for d is substituted in (14–13), we find

$$p_1 = 1 + P_1 d_0 + tP_1{}^2 P_2 P_3$$

$$p_2 = 1 + P_2 d_0 + tP_1 P_2{}^2 P_3 \quad t = 0, 1, 2, \cdots$$

$$p_3 = 1 + P_3 d_0 + tP_1 P_2 P_3{}^2 \qquad (14\text{–}16)$$

To make the product of these numbers an F number, only one condition remains to be fulfilled: they must all be primes. When the successive values of t are introduced and the character of the resulting three numbers checked against a table of primes, it is usually possible to derive a large set of numbers with the Fermat property.

Example.

The simplest example is

$$P_1 = 1, \qquad P_2 = 2, \qquad P_3 = 3$$

and the congruence (14–14) becomes

$$11d \equiv -6 \ (\mathrm{mod}\ 6)$$

with the smallest positive solution $d_0 = 6$. These numerical values, when substituted in the formulas (14–16) give

$$p_1 = 7 + 6t, \qquad p_2 = 13 + 12t, \qquad p_3 = 19 + 18t \qquad (14\text{–}17)$$

as the possible expressions for the three primes defining the F numbers. For $t = 0$ we find the first F number of this type

$$n = 7 \cdot 13 \cdot 19 = 1{,}729$$

an example to which we have already referred.

The first few values of t, such that all three numbers (14–17) are primes, have been tabulated below.

t	p_1	p_2	p_3
0	7	13	19
5	37	73	109
34	211	421	631
44	271	541	811
50	307	613	919
54	331	661	991
55	337	673	1,009
. . . .	. . . .	. . . . .	. . . . .
1,514	9,091	18,181	27,271

When t is limited so that the least prime p_1 does not exceed 10,000, one finds 45 F numbers of the type (14–17).

Example.

In a second example we take

$$P_1 = 1, \qquad P_2 = 2, \qquad P_3 = 5$$

The congruence (14–14) becomes

$$17d \equiv -8 \pmod{10}$$

and its least positive solution is $d_0 = 6$. From (14–16) one deduces the corresponding primes

$$p_1 = 7 + 10t, \qquad p_2 = 13 + 20t, \qquad p_3 = 31 + 50t$$

For $t = 0$, one finds the F number $n = 7 \cdot 13 \cdot 31$.

A similar procedure may be devised to determine the F numbers with four or more prime factors. Let us mention only the example

$$n = p_1 p_2 p_3 p_4$$

where the primes belong to the series

$$p_1 = 7 + 6t, \qquad p_2 = 13 + 12t, \qquad p_3 = 19 + 18t,$$
$$p_4 = 37 + 36t \qquad\qquad (14\text{–}18)$$

The smallest F number of this type

$$n = 7 \cdot 13 \cdot 19 \cdot 37$$

is obtained for $t = 0$, and with the limitation $p_1 < 10,000$ one finds 13 more of them.

As we remarked earlier, the literature on numbers with the Fermat property is scant and there are several natural questions still awaiting solution. It is not even known whether there are infinitely many F numbers, although this seems probable. To illustrate the difficulty, let us take the F numbers that are the product of three primes of the form (14–17). It is known by the theorem of Dirichlet that one can find an unlimited number of values of t for which *one* of the numbers in (14–17), for instance, p_1, becomes a prime. But whether one can find infinitely many values of t such that all three numbers simultaneously become primes is a problem beyond the present power of number theory.

The F numbers by their definition have a basic property in common with the primes, but in other respects they behave quite differently. For instance, one F number may divide another, as in the example

$$n_1 = 7 \cdot 13 \cdot 19, \qquad n_2 = 7 \cdot 13 \cdot 19 \cdot 37$$

or, more generally, any F number with four prime factors (14–18) is divisible by the F number with the three factors (14–17). An F number can even be the product of two other F numbers; for instance,

$$m = (7 \cdot 13 \cdot 19)(37 \cdot 73 \cdot 109)$$

is an example.

Problems.

Find the general form of three prime factors of an F number corresponding to the following values and determine one or more examples in each case.

1. $P_1 = 1,$ $P_2 = 3,$ $P_3 = 4$

2. $P_1 = 1,$ $P_2 = 2,$ $P_3 = 15$

3. $P_1 = 2,$ $P_2 = 3,$ $P_3 = 5$

Bibliography

CARMICHAEL, R. D.: "Note on a New Number Theory Function," *Bulletin American Mathematical Society*, Vol. 16, 232–238 (1910).

———: "On Composite Numbers P Which Satisfy the Fermat Congruence $a^{P-1} \equiv 1 \mod P$," *American Mathematical Monthly*, Vol. 19, 22–27 (1912).

CHERNICK, J.: "On Fermat's Simple Theorem," *Bulletin American Mathematical Society*, Vol. 45, 269–274 (1939).

LEHMER, D. H.: "On the Converse of Fermat's Theorem," *American Mathematical Monthly*, Vol. 43, 347–354 (1936).

POULET, P.: "Table des nombres composés vérificant le théorème de Fermat pour le modul 2 jusqu'à 100,000,000," *Sphinx* (Brussels), Vol. 8, 42–52 (1938).

SISPÁNOV, S.: "Sobre los numeros pseudo-primos," *Boletin matematico*, Vol. 14, 99–106 (1941).

CHAPTER 15

THE CLASSICAL CONSTRUCTION PROBLEMS

15-1. The classical construction problems. Quite early, probably in the fifth century B.C., Greek mathematical investigations led to the study of some geometric construction problems that have remained landmarks in the history of mathematics. Three of these have acquired a particular fame.

1. *The squaring of the circle.* When a circle is given, this problem requires that a square shall be constructed whose area is equal to that of the circle. The same difficulty is involved in the construction of a straight distance equal to the circumference of the circle.

2. *The trisection of the angle.* This problem demands a method for dividing an arbitrary angle into three equal parts.

3. *The doubling of the cube.* This problem is sometimes known as the *Delian problem.* According to tradition, it arose when the Athenians sought the assistance of the oracle at Delos to gain relief from a devastating epidemic. They were advised to double the size of the altar of Apollo, cubical in shape.

These problems enjoyed popular fame among the Greeks; we know, for instance, according to a letter from the mathematician Eratosthenes to King Ptolemy of Egypt, that Euripides mentions the Delian problem in one of his tragedies, now lost. Greek geometers also focused their interest on several other construction problems; we mention especially the construction of *regular polygons*, which we shall discuss in some detail subsequently, and the *Platonic bodies* or regular polyhedrons.

These problems had an inspiring influence and added new

340

aspects to Greek geometry. Special higher curves were introduced and their properties studied, and by such means the trisection of the angle and the duplication of the cube could be accomplished. However, in the strict Greek sense of construction by compass and ruler, the problems remained unsolved in spite of strenuous and ingenious efforts by the Greek and later geometers.

In the seventh and last section of the *Disquisitiones arithmeticae*, Gauss turns to the problem of constructing regular polygons. It may seem out of order to introduce a geometric topic in a work on number theory; hence Gauss feels obliged to explain: "The reader may be surprised at encountering such an investigation which at first view appears wholly dissimilar to it; but the exposition will show very clearly the actual relation between this topic and the transcendental arithmetic."

We shall present a brief account of Gauss's principal results on the construction of regular polygons. For this purpose it is necessary to touch upon some of the principles of geometric construction in general. When it is required that a construction shall be performed by compass and ruler, it is assumed that each of these two instruments shall be used only for a single, specific operation:

1. With the compass, circles with given center and radius can be traced.
2. With the ruler, a straight line can be drawn through two given points.

In these statements it is tacitly included that one can draw circles with arbitrarily large radii and that straight lines can be prolonged indefinitely. Any points or lengths one can deduce from given geometric quantities by a finite number of these two operations are said to have been constructed by compass and ruler.

It is not permissible to apply the two instruments in any other way; for instance, markings on a ruler cannot be utilized. There exists, for instance, a very simple solution of the trisection problem by means of a ruler with two fixed marks.

While we are on the subject of geometric constructions, let us take a small step out of the direct path and mention the rather interesting fact that any construction that can be performed by compass and ruler can be made by compass alone, and also, if a fixed circle has been drawn, the constructions may be achieved by ruler only.

After the basic rules for the construction by means of compass and ruler have been clarified, the next move in the analysis of the construction problems consists in bringing them in relation to the theory of algebraic equations. It must be emph. .sized that the subsequent presentation is essentially expository and that some of the most important steps can only be stated here without any attempts to give proofs.

To determine which quantities can be constructed, let us assume that one performs the geometric operations within a coordinate system in the plane and examine the algebraic operations involved in each step. When two points are given by means of their coordinates, the coefficients of the equation of the straight line passing through them can be computed rationally from the coordinates; dually, when the coefficients of the two straight lines are known, the coordinates of their intersection point can be determined rationally from them. The calculation of the intersection points of a circle and a straight line, or of a circle with another circle, leads to a second-degree equation. The coordinates, therefore, are obtained as the sum of a rational expression in the known coefficients of the equations and the square root of such an expression. The distance between two points is also expressible as a square root.

Since all other constructions can be composed of a series of these simple operations, we conclude from our observations that those magnitudes that can be constructed from given ones may be computed algebraically by repeated applications of the four arithmetic operations and by extracting square roots. But the converse is also true. When a and b are two given lengths, one obtains the distances $a \pm b$ as well as ab and a/b by elementary constructions while the square root $\sqrt{a}$ is the result of taking the mean proportional of 1 and a.

To summarize: The geometric quantities that are constructible from known data by means of compass and ruler correspond algebraically to those expressions that may be deduced from given numbers by repeated use of the four rational operations and square root extraction.

Through this analysis we have succeeded in transferring the construction problems to questions in the theory of equations, since it is relatively easy to show that each constructible expression is the root of an algebraic equation whose coefficients are rational in the given quantities. One way of finding such an equation is the following: Our constructible expression contains a number of square roots. Related expressions can be deduced from it by changing the $\pm$ signs in front of each of these radicals in all possible ways. The equation whose roots are all these quantities is an equation of the desired kind. For instance, the quantity $x = a\sqrt{3}$ satisfies the equation

$$x^2 - 3a^2 = 0$$

The expression

$$x = 1 + \sqrt{3 - \sqrt{5}}$$

is a root of the equation of fourth degree

$$[(x - 1)^2 - 3]^2 - 5 = x^4 - 4x^3 + 8x - 1 = 0$$

This transformation makes it clear that to decide on the possibility of solving a construction problem, one must examine first whether the quantity to be found satisfies an algebraic equation, and second whether this equation has a constructible solution, or, as one prefers to say in equation theory, whether it is solvable by square roots.

The further problem of solving an equation by square roots or, more generally, by radical expressions could not be tackled until the discovery by the two young geniuses, N. H. Abel (1802–1829), Norwegian, and E. Galois (1811–1832), French, of the principles underlying the solution of algebraic equations. It is impossible to discuss these theories here; we will only say that they have been fundamental in the history of the newer phases of algebra and gave rise to the all-pervading mathematical concept of groups. For our purpose a few very simple facts will suffice.

We have seen that any constructible expression satisfies an algebraic equation with coefficients that are rational in the given quantities. There may be several such equations, but among them there is one of minimal degree, which cannot be factored further with rational coefficients, and it divides all other equations of the same kind. From the theory of Galois it follows that for this minimal equation to be solvable by square roots it must have very special properties. One of these is that *its degree must be a power of two*.

We return to the three classical construction problems, and consider first the duplication of the cube. The given cube may have the side a and the doubled cube the side b. Since the volume of one cube is to be the double of the other, they must fulfill the condition

$$b^3 = 2a^3$$

or

$$b = \sqrt[3]{2} \cdot a$$

Therefore, the problem is essentially to construct the number $x = \sqrt[3]{2}$, which is a root of the equation

$$x^3 - 2 = 0$$

This equation cannot be factored into rational factors, and since its degree is not a power of 2, we conclude that a cube cannot be doubled by means of a construction with compass and ruler.

The impossibility of a general construction for the trisection of the angle can be deduced by a similar argument. An angle α can be constructed when one knows $\cos \alpha$ (or $\sin \alpha$) because α occurs in the right triangle with the hypotenuse 1 and one leg equal to $\cos \alpha$. Conversely, when an angle is given, its cosine or sine can be constructed. The problem of trisection of an angle may therefore be expressed: It is required to construct the number $x = \cos \dfrac{\alpha}{3}$ when the number $a = \cos \alpha$ is known. By means of the elementary trigonometric formula

$$\cos 3\theta = 4 \cos^3 \theta - 3 \cos \theta$$

one finds

$$\cos \alpha = 4 \left(\cos \frac{\alpha}{3} \right)^3 - 3 \cos \frac{\alpha}{3}$$

and this may be rewritten as a cubic equation

$$4x^3 - 3x - a = 0$$

In general, one cannot decompose this equation further into factors whose coefficients depend rationally on a; there are numerous values of a, in fact infinitely many rational values of a, within any interval such that the equation cannot be factored rationally. Since the equation is cubic, we conclude that a general construction for trisecting an angle by compass and ruler cannot be found.

The quadrature of the circle is a problem on a different level of difficulty. It is equivalent to finding a construction for the number π, the proportion between the circumference and the diameter of a circle. There is no algebraic equation that is naturally associated with this problem, and the final result is actually to the effect that not only is π not constructible but it is a *transcendental number*, *i.e.*, not the root of any algebraic equation with rational coefficients. The proof was found in 1882 by the German mathematician F. Lindemann, and it was based on methods devised previously by the French mathematician C. Hermite, who, in 1873, showed that the number e, the base of the natural logarithms, is a transcendental number.

The detailed proofs of the impossibility of constructing solutions to the three classical problems leave nothing to be desired in regard to mathematical stringency. Nevertheless, every mathematician has received and undoubtedly will continue to receive new and ingenious constructions purporting to be exact solutions. Usually they have been tested by the inventor on large-scale drawings and the proof of the pudding lies in the eating: no perceptible error has been found. All of these constructions are, needless to say, approximations with errors that are small but definite and computable by elementary trigonometry. Some of the published constructions are of interest due to their simplicity and great accuracy. Any one of them can be improved upon by further complications. Mathematically, an accuracy that leaves no errors observable by the naked eye is not impressive. For instance, π would be constructible if one were permitted to cut off its expansion after the first hundred or first thousand decimals; even by trial and error, the stage of error not perceptible to the eye is reached in a few steps.

15–2. The construction of regular polygons. We now come to the principal topic of this chapter, Gauss's investigations on the construction of regular polygons. A *regular polygon* with n sides has its vertices equidistant on a circle. The size of the circle is unessential so that we shall assume that its radius is $r = 1$. Since each of the sides of the polygon corresponds to a central angle

$$\frac{360°}{n} = \frac{2\pi}{n}$$

the problem is to divide a full angle of 360° into n equal parts.

Any angle can be bisected, so that when a regular polygon with n sides has been obtained, one can successively construct one with $2n, 4n, \ldots$, in general, with $2^\mu n$ sides. On the other hand, from a polygon with $2n$ sides, one can draw one with n sides by joining every second vertex by a side. Consequently, if one so desires, one can limit the considerations only to regular polygons with an odd number of sides. From the fact that regular polygons with 3, 4, and 5 sides can be constructed, it follows that all polygons with

$$2^\mu, \quad 3 \cdot 2^\mu, \quad 5 \cdot 2^\mu$$

sides are obtainable.

It is evident that if one has a polygon with n sides and a is a divisor of n, say, $n = ab$, a polygon with a sides can be derived by taking every bth vertex. More interesting is the fact that the basic result on linear indeterminate equations under certain circumstances permits us to proceed the other way and obtain polygons with a larger number of sides.

From polygons with a and with b sides, where a and b are relatively prime, a polygon with ab sides is obtainable.

To prove the statement we recall that one can find such integers x and y that

$$ax - by = 1$$

Division by ab gives

$$\frac{1}{ab} = \frac{x}{b} - \frac{y}{a}$$

or

$$\frac{360°}{ab} = x\frac{360°}{b} - y\frac{360°}{a}$$

This shows that the central angle for a polygon with ab sides is the difference between two multiples of the central angles of the polygons with a and b sides. For instance, from the polygons with 3 and 5 sides, a polygon with 15 sides is constructible. One concludes also that it would suffice to study the construction of polygons for which the number of sides is an odd prime power.

The construction of a polygon with n sides, or equivalently, an angle $2\pi/n$, may be achieved by using one of the trigonometric functions of the angle, for instance,

$$\cos\frac{2\pi}{n} \quad \text{or} \quad \sin\frac{2\pi}{n} \tag{15-1}$$

By means of the law of cosines one finds the expression

$$s_n = \sqrt{2 - 2\cos\frac{2\pi}{n}}$$

for the side of the polygon. Instead of dealing with these quantities directly, Gauss takes a step that at the time was an innovation: Imaginary or complex numbers are introduced to solve a problem that essentially concerns real quantities.

Let us recall briefly a few properties of complex numbers. Any such number can be written

$$a + ib = r(\cos\varphi + i\sin\varphi), \qquad i = \sqrt{-1} \qquad (15\text{-}2)$$

where a and b are the coordinates in the complex plane, r the *radius vector* or *absolute value*, while φ is the angle or *amplitude*

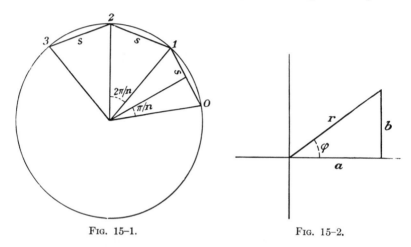

Fig. 15-1.   Fig. 15-2.

that the radius vector makes with the real axis. A complex number (15-2) is multiplied by another

$$a_1 + ib_1 \equiv r_1(\cos\varphi_1 + i\sin\varphi_1)$$

according to the rule

$$(a + ib)(a_1 + ib_1) = rr_1[\cos(\varphi + \varphi_1) + i\sin(\varphi + \varphi_1)]$$

When this result is applied to a product of n equal factors (15-2), one derives the formula

$$(a + ib)^n = r^n(\cos n\varphi + i\sin n\varphi) \qquad (15\text{-}3)$$

known as the *theorem of de Moivre*.

Gauss assumes that a circle with radius 1 and center at the zero point has been drawn in the complex plane. In this circle he inscribes a regular polygon with n sides such that one vertex lies on the positive real axis at the point $x = 1$ (see Fig. 15-1). Then the next vertex will correspond to the complex number

$$\epsilon = \cos\frac{2\pi}{n} + i\sin\frac{2\pi}{n} \qquad (15\text{--}4)$$

and the subsequent ones to

$$\epsilon_2 = \cos\frac{4\pi}{n} + i\sin\frac{4\pi}{n}, \quad \cdots,$$

$$\epsilon_{n-1} = \cos\frac{2\pi}{n}(n-1) + i\sin\frac{2\pi}{n}(n-1) \quad (15\text{--}5)$$

The theorem of de Moivre shows that these numbers are powers of ϵ

$$\epsilon_0 = \epsilon^0 = 1, \qquad \epsilon_1 = \epsilon, \qquad \epsilon_2 = \epsilon^2, \cdots, \epsilon_{n-1} = \epsilon^{n-1} \quad (15\text{--}6)$$

From the same formula (15–3) we conclude further

$$\epsilon^n = \left(\cos\frac{2\pi}{n} + i\sin\frac{2\pi}{n}\right)^n = \cos 2\pi + i\sin 2\pi = 1$$

so that

$$\epsilon^n = 1$$

This result establishes that ϵ as well as all its powers (15–6) must be roots of the algebraic equation

$$x^n - 1 = 0 \qquad (15\text{--}7)$$

For this reason one calls the roots (15–6) the nth *roots of unity*, while the equation (15–7) is known as the *equation of the division of the circle* or the *cyclotomic equation*.

The two trigonometric functions (15–1), on which the construction of the regular polygon depends, occur as the components of the nth roots of unity (15–4). When they can be expressed by square root operations, the same is true for ϵ, recalling that $i = \sqrt{-1}$. Therefore, if the regular polygon with n sides can be constructed with compass and ruler, the corresponding cyclotomic equation (15–7) can be solved by square roots.

Equation (15–7) is not the equation of minimal degree that the nth root of unity ϵ satisfies, since it can be factored rationally;

for instance, $x - 1$ is an obvious factor. In general, the roots (15–6) of the cyclotomic equation fall into two groups. Some of them cannot be roots of unity for a smaller exponent than n; these are usually called the *primitive roots*. Others do satisfy equations of the type (15–7) with lower exponents and these are the *nonprimitive roots*. We may remark that these primitive roots constitute a different, although related, concept from the primitive roots we introduced previously when studying the residue classes of the integers for some modul.

It is simple to decide when a root of unity

$$\epsilon_k = \epsilon^k = \cos\frac{2\pi k}{n} + i\sin\frac{2\pi k}{n} \qquad (15\text{–}8)$$

is primitive. If it should satisfy an equation

$$x^t = 1, \qquad t < n$$

the theorem of de Moivre shows that

$$\epsilon_k{}^t = \cos\frac{2\pi kt}{n} + i\sin\frac{2\pi kt}{n} = 1$$

This is only possible if the amplitude

$$2\pi\frac{kt}{n}$$

is an integral multiple of 2π; in other words the number kt must be divisible by n. When k is relatively prime to n, the smallest t that will satisfy this condition is $t = n$, while a smaller t can be found when k and n have a common factor. Thus we have:

An nth root of unity ϵ_k, defined in (15–8), is primitive only when k is relatively prime to n.

Since the roots in (15–6) that are primitive correspond to those numbers k that are less than and relatively prime to n, we can state further:

The number of primitive nth roots of unity is equal to $\varphi(n)$, where φ denotes Euler's function.

The subsequent step in the algebra of the roots of unity is to demonstrate that the $\varphi(n)$ primitive roots satisfy an equation with rational coefficients of degree $\varphi(n)$, and that this equation cannot be factored further rationally. Again we must abstain from giving a proof. To this minimal equation for the primitive roots, we apply our previous criterion limiting the degree of an equation solvable by square roots. This produces the interesting result:

For the equation of the nth roots of unity to be solvable by means of square roots, it is necessary that $\varphi(n)$ be a power of 2.

This places a strong restriction on the number n. To analyze its implications, let

$$n = 2^{\alpha_0} p_1^{\alpha_1} \cdots p_r^{\alpha_r} \tag{15-9}$$

be the prime factorization of n. The number

$$\varphi(n) = 2^{\alpha_0-1} p_1^{\alpha_1-1}(p_1 - 1) \cdots p_r^{\alpha_r-1}(p_r - 1)$$

can be a power of 2 only when each of its factors is such a power. One concludes first that none of the odd prime factors p_i can occur, so that all exponents $\alpha_1, \ldots, \alpha_r$ in (15-9) must be equal to 1. Second, the numbers $p_i - 1$ are powers of 2; hence the odd primes dividing n are of the form

$$p_i = 2^{k_i} + 1 \tag{15-10}$$

But these are actually the Fermat primes, which were examined in Chap. 4. We found that such a number as (15-10) cannot be a prime except when the exponent is itself a power of 2 and that the Fermat primes, therefore, are defined by an expression

$$F_t = 2^{2^t} + 1$$

We mentioned also that so far the study of these numbers has revealed only five Fermat primes, namely,

$$F_0 = 3, \qquad F_1 = 5, \qquad F_2 = 17, \qquad F_3 = 257, \qquad F_4 = 65,537$$

Through these observations we have arrived at Gauss's fundamental result:

A regular polygon with n sides can be constructed by compass and ruler only when the number n is of the form

$$n = 2^\alpha p_1 p_2 \cdots p_r$$

where the prime factors are Fermat primes.

Our previous discussion has been directed towards showing that this condition is necessary. Gauss proves conversely that it is also a sufficient one, by demonstrating that a polygon with p sides can be constructed when p is a Fermat prime. In this case he finds that the equation for the primitive pth roots of unity can be solved by a series of second-degree equations. We shall not go through the details of the general proof, but only consider a couple of examples sufficient to clarify the underlying principles.

Problem.

Find all polygons with less than 100 sides that can be constructed with compass and ruler.

15–3. Examples of constructible polygons. When p is a prime, the number of primitive pth roots of unity must be $\varphi(p) = p - 1$, and clearly the only nonprimitive root is $x = 1$. Since they all satisfy the equation

$$x^p - 1 = 0$$

the primitive ones are the roots of

$$\frac{x^p - 1}{x - 1} = x^{p-1} + x^{p-2} + \cdots + x^2 + x + 1 = 0 \quad (15\text{–}11)$$

These roots, as we mentioned, are all some power of

$$\epsilon = \cos \frac{2\pi}{p} + i \sin \frac{2\pi}{p}$$

We notice further that the two roots

$$\epsilon^k = \cos \frac{2\pi}{p} k + i \sin \frac{2\pi}{p} k$$

$$\epsilon^{p-k} = \epsilon^{-k} = \cos \frac{2\pi}{p} k - i \sin \frac{2\pi}{p} k$$

are conjugate imaginary, and their sum

$$\epsilon^k + \epsilon^{-k} = 2 \cos \frac{2\pi}{p} k$$

is real.   In particular

$$\eta = \epsilon + \epsilon^{-1} = 2 \cos \frac{2\pi}{p} > 0 \qquad (15\text{--}12)$$

and this number may serve to construct the polygon.

For the smallest Fermat prime $p = 3$, (15–11) becomes

$$x^2 + x + 1 = 0$$

We substitute $x = \epsilon$ and obtain after division by ϵ

$$\eta + 1 = \epsilon + \epsilon^{-1} + 1 = 0$$

According to (15–12), this gives

$$\cos \frac{2\pi}{3} \equiv -\frac{1}{2}$$

and for the side of the polygon, one finds the value

$$s_3 = \sqrt{3}$$

The next Fermat prime is $p = 5$.   Here ϵ satisfies the equation

$$\epsilon^4 + \epsilon^3 + \epsilon^2 + \epsilon + 1 = 0$$

and division by ϵ^2 yields

$$\epsilon^2 + \epsilon^{-2} + \epsilon + \epsilon^{-1} + 1 = 0 \qquad (15\text{--}13)$$

From

$$\eta = \epsilon + \epsilon^{-1}$$

we obtain, by squaring,

$$\eta^2 - 2 = \epsilon^2 + \epsilon^{-2}$$

When these values are substituted into (15–13), it follows that η is the root of the second-degree equation

$$\eta^2 + \eta - 1 = 0$$

The solution of this equation is

$$\eta = \frac{\sqrt{5} - 1}{2}$$

where we have taken the plus sign for the square root since η is positive according to (15–12). We obtain further

$$\cos \frac{2\pi}{5} = \frac{\sqrt{5} - 1}{4}$$

and the side of the pentagon is computed to be

$$s_5 = \tfrac{1}{2} \sqrt{10 - 2\sqrt{5}}$$

In our last example we take the Fermat prime $p = 17$, and in this case the principles of the general theory emerge more clearly. When the number 0 is excluded, the sixteen other remainders (mod 17) may be written

$$\pm 1, \quad \pm 2, \quad \pm 3, \quad \pm 4, \quad \pm 5, \quad \pm 6, \quad \pm 7, \quad \pm 8 \quad (15\text{–}14)$$

We shall first divide these numbers into two classes

$$\pm 1, \quad \pm 2, \quad \pm 4, \quad \pm 8 \qquad (15\text{–}15a)$$
$$\pm 3, \quad \pm 5, \quad \pm 6, \quad \pm 7 \qquad (15\text{–}15b)$$

The numbers in (15–15a) are known as the *quadratic residues* (mod 17); they are obtained by squaring the numbers in (15–14) and taking the remainders (mod 17). The remaining numbers in (15–14), which have been put in the set (15–15b) are the *quadratic nonresidues*. One should notice that the numbers in (15–15b) can be derived from those in (15–15a) by multiplication with some nonresidue, for instance, 3.

Second, the remainders (15–14) shall be distributed into four classes, each of four numbers

$$\pm 1, \quad \pm 4 \qquad (15\text{–}16a)$$
$$\pm 2, \quad \pm 8 \qquad (15\text{–}16b)$$
$$\pm 3, \quad \pm 5 \qquad (15\text{–}16c)$$
$$\pm 6, \quad \pm 7 \qquad (15\text{–}16d)$$

Here the first set consists of the remainders of the fourth powers of the numbers in (15–14) or, equivalently, of the squares of the numbers in (15–15a). They are called the *biquadratic residues*. To obtain (15–16b), the numbers in (15–16a) are multiplied by some number in (15–15a) not already in (15–16a), for instance, 2. The numbers in (15–16c) follow from (15–16a) through multiplication by some number not in the two preceding groups, and (15–16d) is derived similarly.

Through a third division the remainders (15–14) fall into eight sets of two numbers. Here we use the basic group ± 1, namely, the residues of the eighth powers of the numbers in (15–14). By this process the first set (15–16a) splits into ± 1 and ± 4, and the other sets in (15–16) are divided similarly.

After these preliminaries we turn to the solution of (15–11) for $p = 17$. When the root $x = \epsilon$ is substituted and the equation divided by ϵ^8, it follows that

$$\epsilon + \epsilon^{-1} + \epsilon^2 + \epsilon^{-2} + \cdots + \epsilon^8 + \epsilon^{-8} = -1 \quad (15\text{–}17)$$

At this stage Gauss introduces two quantities which he calls the *first periods*

$$\left.\begin{array}{l} \rho = \epsilon + \epsilon^{-1} + \epsilon^2 + \epsilon^{-2} + \epsilon^4 + \epsilon^{-4} + \epsilon^8 + \epsilon^{-8} \\ \rho_1 = \epsilon^3 + \epsilon^{-3} + \epsilon^5 + \epsilon^{-5} + \epsilon^6 + \epsilon^{-6} + \epsilon^7 + \epsilon^{-7} \end{array}\right\} \quad (15\text{–}18)$$

These periods, as one sees, are the sums of the roots whose exponents are the numbers in the two classes (15–15a) and (15–15b). Both periods are real, for instance

$$\rho = 2\cos\frac{2\pi}{17} + 2\cos\frac{4\pi}{17} + 2\cos\frac{8\pi}{17} + 2\cos\frac{16\pi}{17}$$

and it is readily checked that ρ is positive and ρ_1 negative.

The periods (15–18) are the roots of an equation of second degree

$$r(x) = (x - \rho)(x - \rho_1) = x^2 - (\rho + \rho_1)x + \rho\rho_1 \quad (15\text{–}19)$$

which we shall show has rational coefficients. From (15–17) and (15–18) one concludes immediately

$$\rho + \rho_1 = -1$$

The computation of the product of the two periods (15–18) is somewhat more cumbersome. By direct multiplication one obtains $8 \cdot 8 = 64$ products, each a power of ϵ. One verifies that every term ϵ^k with $k \neq 0$ occurs equally often and, since there are 16 different powers, each of them appears four times. According to (15–17) we conclude that

$$\rho\rho_1 = -4$$

These investigations show that the equation (15–19) has the form

$$x^2 + x - 4 = 0$$

and its roots are

$$\rho = \frac{\sqrt{17} - 1}{2}, \qquad \rho_1 = \frac{-\sqrt{17} - 1}{2} \qquad (15\text{–}20)$$

In the next step we study the *second periods*

$$\left.\begin{aligned}
\sigma &= \epsilon + \epsilon^{-1} + \epsilon^4 + \epsilon^{-4} \\
\sigma_1 &= \epsilon^2 + \epsilon^{-2} + \epsilon^8 + \epsilon^{-8} \\
\sigma_2 &= \epsilon^3 + \epsilon^{-3} + \epsilon^5 + \epsilon^{-5} \\
\sigma_3 &= \epsilon^6 + \epsilon^{-6} + \epsilon^7 + \epsilon^{-7}
\end{aligned}\right\} \qquad (15\text{–}21)$$

Here the various sets (15–16) serve as the exponents for the terms in periods. The periods are all real, for instance,

$$\sigma = \cos\frac{2\pi}{17} + \cos\frac{8\pi}{17} > 0$$

Similarly one finds that σ_2 is positive while σ_1 and σ_3 are negative.

The four periods (15–21) in pairs, σ, σ_1 and σ_2, σ_3, are roots of second-degree equations whose coefficients can be expressed by the first periods. From (15–21) and (15–18) one sees that

$$\sigma + \sigma_1 = \rho, \qquad \sigma_2 + \sigma_3 = \rho_1$$

and by multiplication from (15–17)

$$\sigma\sigma_1 = -1, \qquad \sigma_2\sigma_3 = -1$$

so that the two quadratic equations are

$$x^2 - \rho x - 1 = 0, \qquad x^2 - \rho_1 x - 1 = 0$$

For the solution of these equations, one finds by our previous remark about the signs of the periods

$$\sigma = \frac{\rho + \sqrt{\rho^2 + 4}}{2}, \qquad \sigma_1 = \frac{\rho - \sqrt{\rho^2 + 4}}{2}$$

$$\sigma_2 = \frac{\rho_1 + \sqrt{\rho_1^2 + 4}}{2}, \qquad \sigma_3 = \frac{\rho_1 - \sqrt{\rho_1^2 + 4}}{2}$$

Here one can substitute the values (15–20) for ρ and ρ_1 to obtain the explicit expressions for the second periods; for instance,

$$\sigma = \tfrac{1}{4}(\sqrt{17} - 1 + \sqrt{34 - 2\sqrt{17}}),$$
$$\sigma_2 = \tfrac{1}{4}(-\sqrt{17} - 1 + \sqrt{34 + 2\sqrt{17}}) \quad (15\text{–}22)$$

Finally the third-order periods should be computed. We shall need only two of them, namely,

$$\eta = \epsilon + \epsilon^{-1} = 2\cos\frac{2\pi}{17}, \qquad \eta_1 = \epsilon^4 + \epsilon^{-4} = 2\cos\frac{8\pi}{17} \quad (15\text{–}23)$$

where the exponents are taken from (15–16a). Again these two quantities satisfy a second-degree equation whose coefficients can be expressed by means of the second periods. From (15–23) one obtains

$$\eta + \eta_1 = \sigma, \qquad \eta\eta_1 = \sigma_2$$

and the equation is

$$x^2 - \sigma x + \sigma_2 = 0$$

The expressions (15–23) show that $\eta > \eta_1$, and therefore the solution of the quadratic equation gives us

$$\eta = 2\cos\frac{2\pi}{17} = \frac{\sigma + \sqrt{\sigma^2 - 4\sigma_2}}{2}$$

Here the values (15–22) may be substituted, and the final formula in terms of square roots will emerge. The reader may compute the length of the side s_{17} of the regular polygon with 17 sides. Various fairly simple methods of construction have been devised.

After having completed these investigations, towards the end of the *Disquisitiones*, the young Gauss states with justifiable pride:

> There is certainly good reason to be astonished that while the division of the circle in 3 and 5 parts having been known already at the time of *Euclid*, one has added nothing to these discoveries in a period of two thousand years and that all geometers have considered it certain that, except for these divisions and those which may be derived from them (divisions into 2^μ, 15, $3 \cdot 2^\mu$, $5 \cdot 2^\mu$, $15 \cdot 2^\mu$ parts), one could not achieve any others by geometric constructions.

It has been told that Gauss proposed, perhaps not too seriously, that a polygon with 17 sides be inscribed on his grave, emulating the tombstone of Archimedes, which was decorated by a figure of a sphere and the circumscribed cylinder, suggesting his formula for the area of a sphere. On Gauss's simple grave in Göttingen there is no such polygon, but it does appear on the monument in his native town of Brunswick.

Gauss's results on the construction of regular polygons by compass and ruler represent a great achievement, but the final solution of the problem is not yet in sight. Gauss transfers the whole question to number theory, to the determination of the Fermat primes. Whether there exist any others than the five we have mentioned, no one knows. It is possible that the new electronic computing devices may be of assistance in discovering others. But the general problems, for instance, the question whether there might be an infinite number of Fermat primes, lie beyond the reach of the present methods of number theory.

Bibliography

DICKSON, L. E.: *Modern Algebraic Theories*, Benj. H. Sanborn & Co., New York, 1930.

MacDUFFEE, C. C.: *An Introduction to Abstract Algebra*, John Wiley & Sons, Inc., New York, 1940.

ROUSE-BALL, W. W.: *Mathematical Recreations and Essays*. Revised by H. S. M. Coxeter, eleventh edition, The Macmillan Company, New York, 1939.

THOMAS, J. M.: *Theory of Equations*, McGraw-Hill Book Company, Inc., New York, 1938.

SUPPLEMENT

RECENT NUMERICAL RESULTS

The predictions about the usefulness of the electronic computers for calculations in number theory have been amply fulfilled since the manuscript for this book was first prepared. Several interesting computations, vastly beyond the range of ordinary calculating machines have been carried out by means of various types of electronic computers and more will undoubtedly follow. Since such studies are not income-producing tasks many have been achieved in the hours of the night where the machines would otherwise have been idle. All have been repeated and checked, sometimes on different computers with new programming and other operators, since men and machines tend to fall into errors of habit and constitution.

Mersenne primes. Among the most formidable of these calculations is the search for new Mersenne primes

$$M_p = 2^p - 1$$

Professor D. H. Lehmer has made a wide sweep for possible primes M_p and he announced in 1952 and 1953 that the values corresponding to

$$p = 521, 607, 1{,}279, 2{,}203, 2{,}281$$

are primes. For the last two of these Mersenne primes, after coding and preparations, the actual running time of the SWAC calculator amounted to 59 and 66 minutes, respectively. In comparison with these giants the Mersenne prime M_{127} given on page 73 appears quite puny. The Mersenne prime $M_{2,281}$ is a number with 687 digits. Professor H. S. Uhler has taken the trouble of calculating explicitly the new perfect numbers corresponding to these primes.

Various conjectures have been made about Mersenne primes, but as far as they have been checked none of them seems to have any validity. It has been noticed, for instance, that when the Mersenne primes

$$p = 3, 7, 31, 127$$

are used as exponents in M_p they give new Mersenne primes. This is no general rule, however, since Professor R. M. Robinson recently announced that calculations carried out on an ILLIAC machine by D. J. Wheeler show that for the Mersenne prime $p = 8,191$ the corresponding M_p is not a prime.

In this connection let us also mention that the Fermat numbers

$$F_n = 2^{2^n} + 1$$

have been further examined, but no new primes have been found. One of the latest results is an actual factor of F_{10}.

Bibliography

UHLER, H. S.: "A brief history of the investigations on Mersenne numbers and the latest immense primes," *Scripta Mathematica*, Vol. 18, 122–131 (1952).

UHLER, H. S.: "On the 16th and 17th perfect numbers," *Scripta Mathematica*, Vol. 19, 128–131 (1953).

BANG, T.: "Store primtal (Large primes)," *Nordisk Matematisk Tidskrift*, Vol. 2, 157–168 (1954).

ROBINSON, R. M.: "Mersenne and Fermat numbers," *Proceedings of the American Mathematical Society*, Vol. 5, 842–846 (1954).

Odd perfect numbers. A great variety of results have been obtained on the possible forms of odd perfect numbers, particularly by A. Brauer and H. J. Kanold. It has also been shown by Kanold [*Journal für die reine und angewandte Mathematik*, Vol. 186, 25–29 (1944)] that there are no odd perfect numbers below 1.4×10^{14}. One of my students, J. B. Muskat, has informed me that he has been able to raise this bound to 10^{18}.

GENERALIZATIONS OF THE PERFECT NUMBERS

New lists of *multiply perfect numbers* have recently been published by B. Franqui and M. García [*American Mathematical*

Monthly, Vol. 60, 169–171 (1953)] and also by A. L. Brown [*Scripta Mathematica*, Vol. 20, 103–106 (1954)] adding more than 200 such numbers to those previously known. In this connection let us also mention a list by P. Poulet [*Scripta Mathematica*, Vol. 14, 77 (1948)] over new couples of *amicable numbers*.

Another generalization of the perfect numbers has been suggested by the author. It is not difficult to prove that for a perfect number n the harmonic mean $H(n)$ of the divisors of n as defined in Sec. 5.1 is always an integer, while for other numbers this is only rarely the case. Thus the integers with integral harmonic mean for the divisors may be considered a generalization of the perfect numbers. They seem to share the property that they are all even. This has been checked by M. García for all such numbers up to 10,000,000.

Bibliography

ORE, O.: "On the averages of the divisors of a number," *American Mathematical Monthly*, Vol. 55, 615–619 (1948).

GARCÍA, M.: "On numbers with integral harmonic mean," *American Mathematical Monthly*, Vol. 61, 89–96 (1954).

Prime tables. Based on calculations by Kulik, Poletti, and Porter a new prime table covering the 11th million has been published by N. G. Beeger (Amsterdam, 1951). There are 61,938 primes in the 11th million. It may also be mentioned that lists of prime twins up to 200,000 have been prepared by E. S. Selmer and G. Nesheim [*Det Kongelige Norske Videnskabers Selskabs, Forhandlinger Trondheim*, Vol. 15, 95–98 (1942)] and up to 300,000 by H. Tietze [*Sitzungsberichte der mathematisch-naturwissenschaftlichen Klasse der bayerischen Akadamie der Wissenschaften zu München*, 57–72 (1947)].

We indicated in Sec. 14.1 how tables of the composite numbers n, satisfying the congruence

$$2^{n-1} \equiv 1 (\text{mod } n) \tag{1}$$

were an essential aid to the factorization of large numbers. D. H. Lehmer [*American Mathematical Monthly*, Vol. 56, 300–309

(1949)] has extended these tables to $n \doteq 200{,}000{,}000$ by means of the Army Ordnance ENIAC computer. Sierpinsky pointed out that there is an infinity of composite numbers satisfying (1), and later Lehmer and P. Erdös proved that there is an infinity of such numbers with any given number of prime factors [*American Mathematical Monthly*, Vol. 56, 623–624, (1949)].

Fermat's problem. The studies on Fermat's problem have been continued quite intensively, both in theory and by means of high-speed computers. Among the most notable contributions are those by Professor H. S. Vandiver and the numerical calculations carried out by D. H. Lehmer, Emma Lehmer, and J. Selfridge at his suggestion. It is shown that the equation

$$x^n + y^n = z^n \qquad n > 2 \qquad (2)$$

can have no nonzero integral solutions for any value of $n < 2{,}521$. In this case the extensive calculations have had the ideal effect of bringing in new points of view also in regard to the theoretical problems. "Thanks to SWAC for special exponents we have really come to grips with the Fermat problem," as Vandiver expresses it [*Proceedings of the National Academy of Science*, Vol. 40, 474–480 (1954)]. (Word has just been received from Professor Vandiver that a series of calculations on Fermat's theorem by C. A. Nicol and J. Selfridge has been completed. It has been verified that Fermat's theorem is true for all exponents up to $n = 4{,}000$.)

A number of improved estimates of the lowest possible values of the numbers x, y, and z in a possible solution of (2) have also been made, particularly by Obláth and Inkeri. On the basis of Vandiver's calculations and new estimating methods by Duparc and Wijngaarden these values could again be raised essentially. However, there does not seem to be much point in continuing along these lines; Obláth observes in connection with his own estimates that x and y would have to exceed a number which at good speed would take more than two centuries to write and a strip of 4,000 miles to print.

GENERAL BIBLIOGRAPHY

For a complete, encyclopedic account of the history of the discoveries in number theory up to 1918 the reader is referred to:

DICKSON, L. E.: "History of the Theory of Numbers," 3 vols., *Carnegie Institution of Washington Publication* 256, 1919–1923.

A review of the existing table material on number-theory questions can be found in:

LEHMER, D. H.: "Guide to the Tables in the Theory of Numbers," *National Research Council Bulletin* 105, Washington, 1941.

A considerable selection of translations and reproductions of essential contributions to number theory is contained in:

SMITH, D. E.: *A Source Book in Mathematics*, McGraw-Hill Book Company, Inc., New York, 1929.

The following is a list of English books on number theory for the reader who wishes to pursue the subject further:

CARMICHAEL, R. D.: *Theory of Numbers*, John Wiley & Sons, Inc., New York, 1914.

————: *Diophantine Analysis*, John Wiley & Sons, Inc., New York, 1915.

DICKSON, L. E.: *Introduction to the Theory of Numbers*, University of Chicago Press, Chicago, 1929.

————: *Studies in the Theory of Numbers*, University of Chicago Press, Chicago, 1930.

————: *Modern Elementary Theory of Numbers*, University of Chicago Press, Chicago, 1939.

HARDY, G. H. and E. M. WRIGHT: *An Introduction to the Theory of Numbers*, Oxford University Press, New York, 1938.

INGHAM, A. E.: *The Distribution of Prime Numbers*, Cambridge University Press, London, 1932.

USPENSKY, J. V., and M. A. HEASLET: *Elementary Number Theory*, McGraw-Hill Book Company, Inc., New York, 1939.

WRIGHT, H. N.: *First Course in Theory of Numbers*, John Wiley & Sons, Inc., New York, 1939.

NAME INDEX

SUBJECT INDEX

A

Abacists, 21, 185
Abacus, 14, 15, 225, 227
Absolute value, 28, 348
Absorption law, 103
Abundant numbers, 94, 95
Al-Fakhri, 185, 187, 193, 208
Algebra (Euler), 60, 126, 128, 132, 138, 141, 206
Algebra, 20, 123
  syncopated, 181
Algebraic congruences, 234, 235, 249–258
Algebraic numbers, 206, 207
Algorism, 20, 21
  (*See also* Euclid's algorism)
Algorismus, 21
Aliquot parts, 86, 91, 98
Al-Jabr wal-Muqabalah, 19, 20, 187
Al-Kafi fil hisab, 185
Amicable numbers, 27, 96–100
Amplitude, 348, 350
Apices, 20
Arabic numerals, 19, 24, 117
Arithmetic (Sun-Tse), 245
Arithmetic, 180
Arithmetic mean, 90, 91
Arithmetic series, 79
Arithmetics (Diophantos), 180–185, 194, 199, 270
Associative law, 48, 103
Astrology, 28

B

Attic numerals, 11, 15
Average, 90

Babylonian numerals and system, 2, 16–18, 36, 37, 172–179, 188, 312, 324
Base (of number systems), 3
Bija-Ganita, 123, 136, 193
Binary number systems, 2, 37
Billion, 5
Book of Precious Things in the Art of Reckoning, The, 139
Brahma-Sphuta-Siddhanta, 122, 193
Brahmi numerals, 19
Bureau, 15

C

Calculations, 14, 15
Calculus, 15, 55
Canon arithmeticus, 285, 301
Casting, on the lines, 15, 225, 227
  out nines, 15, 225, 227, 229–231
Cattle problem, 140
Checks, 15
  numerical, 225–233
Chinese remainder theorem, 240, 246–248, 251, 264, 265, 294, 304
Chinese-Japanese numerals, 11, 12, 16, 17
Cipher, 11, 20
Ciphered numerals, 12–13, 19

A CATALOG OF SELECTED
DOVER BOOKS
IN SCIENCE AND MATHEMATICS

A CATALOG OF SELECTED
DOVER BOOKS
IN SCIENCE AND MATHEMATICS

QUALITATIVE THEORY OF DIFFERENTIAL EQUATIONS, V.V. Nemytskii and V.V. Stepanov. Classic graduate-level text by two prominent Soviet mathematicians covers classical differential equations as well as topological dynamics and ergodic theory. Bibliographies. 523pp. 5⅜ × 8½. 65954-2 Pa. $10.95

MATRICES AND LINEAR ALGEBRA, Hans Schneider and George Phillip Barker. Basic textbook covers theory of matrices and its applications to systems of linear equations and related topics such as determinants, eigenvalues and differential equations. Numerous exercises. 432pp. 5⅜ × 8½. 66014-1 Pa. $9.95

QUANTUM THEORY, David Bohm. This advanced undergraduate-level text presents the quantum theory in terms of qualitative and imaginative concepts, followed by specific applications worked out in mathematical detail. Preface. Index. 655pp. 5⅜ × 8½. 65969-0 Pa. $13.95

ATOMIC PHYSICS (8th edition), Max Born. Nobel laureate's lucid treatment of kinetic theory of gases, elementary particles, nuclear atom, wave-corpuscles, atomic structure and spectral lines, much more. Over 40 appendices, bibliography. 495pp. 5⅜ × 8½. 65984-4 Pa. $12.95

ELECTRONIC STRUCTURE AND THE PROPERTIES OF SOLIDS: The Physics of the Chemical Bond, Walter A. Harrison. Innovative text offers basic understanding of the electronic structure of covalent and ionic solids, simple metals, transition metals and their compounds. Problems. 1980 edition. 582pp. 6⅛ × 9¼. 66021-4 Pa. $14.95

BOUNDARY VALUE PROBLEMS OF HEAT CONDUCTION, M. Necati Özisik. Systematic, comprehensive treatment of modern mathematical methods of solving problems in heat conduction and diffusion. Numerous examples and problems. Selected references. Appendices. 505pp. 5⅜ × 8½. 65990-9 Pa. $11.95

A SHORT HISTORY OF CHEMISTRY (3rd edition), J.R. Partington. Classic exposition explores origins of chemistry, alchemy, early medical chemistry, nature of atmosphere, theory of valency, laws and structure of atomic theory, much more. 428pp. 5⅜ × 8½. (Available in U.S. only) 65977-1 Pa. $10.95

A HISTORY OF ASTRONOMY, A. Pannekoek. Well-balanced, carefully reasoned study covers such topics as Ptolemaic theory, work of Copernicus, Kepler, Newton, Eddington's work on stars, much more. Illustrated. References. 521pp. 5⅜ × 8½. 65994-1 Pa. $12.95

PRINCIPLES OF METEOROLOGICAL ANALYSIS, Walter J. Saucier. Highly respected, abundantly illustrated classic reviews atmospheric variables, hydrostatics, static stability, various analyses (scalar, cross-section, isobaric, isentropic, more). For intermediate meteorology students. 454pp. 6⅛ × 9¼. 65979-8 Pa. $12.95

RELATIVITY, THERMODYNAMICS AND COSMOLOGY, Richard C. Tolman. Landmark study extends thermodynamics to special, general relativity; also applications of relativistic mechanics, thermodynamics to cosmological models. 501pp. 5⅜ × 8½. 65383-8 Pa. $12.95

APPLIED ANALYSIS, Cornelius Lanczos. Classic work on analysis and design of finite processes for approximating solution of analytical problems. Algebraic equations, matrices, harmonic analysis, quadrature methods, much more. 559pp. 5⅜ × 8½. 65656-X Pa. $12.95

SPECIAL RELATIVITY FOR PHYSICISTS, G. Stephenson and C.W. Kilmister. Concise elegant account for nonspecialists. Lorentz transformation, optical and dynamical applications, more. Bibliography. 108pp. 5⅜ × 8½. 65519-9 Pa. $4.95

INTRODUCTION TO ANALYSIS, Maxwell Rosenlicht. Unusually clear, accessible coverage of set theory, real number system, metric spaces, continuous functions, Riemann integration, multiple integrals, more. Wide range of problems. Undergraduate level. Bibliography. 254pp. 5⅜ × 8½. 65038-3 Pa. $7.95

INTRODUCTION TO QUANTUM MECHANICS With Applications to Chemistry, Linus Pauling & E. Bright Wilson, Jr. Classic undergraduate text by Nobel Prize winner applies quantum mechanics to chemical and physical problems. Numerous tables and figures enhance the text. Chapter bibliographies. Appendices. Index. 468pp. 5⅜ × 8½. 64871-0 Pa. $11.95

ASYMPTOTIC EXPANSIONS OF INTEGRALS, Norman Bleistein & Richard A. Handelsman. Best introduction to important field with applications in a variety of scientific disciplines. New preface. Problems. Diagrams. Tables. Bibliography. Index. 448pp. 5⅜ × 8½. 65082-0 Pa. $11.95

MATHEMATICS APPLIED TO CONTINUUM MECHANICS, Lee A. Segel. Analyzes models of fluid flow and solid deformation. For upper-level math, science and engineering students. 608pp. 5⅜ × 8½. 65369-2 Pa. $13.95

ELEMENTS OF REAL ANALYSIS, David A. Sprecher. Classic text covers fundamental concepts, real number system, point sets, functions of a real variable, Fourier series, much more. Over 500 exercises. 352pp. 5⅜ × 8½. 65385-4 Pa. $9.95

PHYSICAL PRINCIPLES OF THE QUANTUM THEORY, Werner Heisenberg. Nobel Laureate discusses quantum theory, uncertainty, wave mechanics, work of Dirac, Schroedinger, Compton, Wilson, Einstein, etc. 184pp. 5⅜ × 8½. 60113-7 Pa. $5.95

INTRODUCTORY REAL ANALYSIS, A.N. Kolmogorov, S.V. Fomin. Translated by Richard A. Silverman. Self-contained, evenly paced introduction to real and functional analysis. Some 350 problems. 403pp. 5⅜ × 8½. 61226-0 Pa. $9.95

PROBLEMS AND SOLUTIONS IN QUANTUM CHEMISTRY AND PHYSICS, Charles S. Johnson, Jr. and Lee G. Pedersen. Unusually varied problems, detailed solutions in coverage of quantum mechanics, wave mechanics, angular momentum, molecular spectroscopy, scattering theory, more. 280 problems plus 139 supplementary exercises. 430pp. 6½ × 9¼. 65236-X Pa. $12.95

ASYMPTOTIC METHODS IN ANALYSIS, N.G. de Bruijn. An inexpensive, comprehensive guide to asymptotic methods—the pioneering work that teaches by explaining worked examples in detail. Index. 224pp. 5⅜ × 8½. 64221-6 Pa. $6.95

OPTICAL RESONANCE AND TWO-LEVEL ATOMS, L. Allen and J.H. Eberly. Clear, comprehensive introduction to basic principles behind all quantum optical resonance phenomena. 53 illustrations. Preface. Index. 256pp. 5⅜ × 8½.
65533-4 Pa. $7.95

COMPLEX VARIABLES, Francis J. Flanigan. Unusual approach, delaying complex algebra till harmonic functions have been analyzed from real variable viewpoint. Includes problems with answers. 364pp. 5⅜ × 8½. 61388-7 Pa. $8.95

ATOMIC SPECTRA AND ATOMIC STRUCTURE, Gerhard Herzberg. One of best introductions; especially for specialist in other fields. Treatment is physical rather than mathematical. 80 illustrations. 257pp. 5⅜ × 8½. 60115-3 Pa. $5.95

APPLIED COMPLEX VARIABLES, John W. Dettman. Step-by-step coverage of fundamentals of analytic function theory—plus lucid exposition of five important applications: Potential Theory; Ordinary Differential Equations; Fourier Transforms; Laplace Transforms; Asymptotic Expansions. 66 figures. Exercises at chapter ends. 512pp. 5⅜ × 8½. 64670-X Pa. $11.95

ULTRASONIC ABSORPTION: An Introduction to the Theory of Sound Absorption and Dispersion in Gases, Liquids and Solids, A.B. Bhatia. Standard reference in the field provides a clear, systematically organized introductory review of fundamental concepts for advanced graduate students, research workers. Numerous diagrams. Bibliography. 440pp. 5⅜ × 8½. 64917-2 Pa. $11.95

UNBOUNDED LINEAR OPERATORS: Theory and Applications, Seymour Goldberg. Classic presents systematic treatment of the theory of unbounded linear operators in normed linear spaces with applications to differential equations. Bibliography. 199pp. 5⅜ × 8½. 64830-3 Pa. $7.95

LIGHT SCATTERING BY SMALL PARTICLES, H.C. van de Hulst. Comprehensive treatment including full range of useful approximation methods for researchers in chemistry, meteorology and astronomy. 44 illustrations. 470pp. 5⅜ × 8½. 64228-3 Pa. $10.95

CONFORMAL MAPPING ON RIEMANN SURFACES, Harvey Cohn. Lucid, insightful book presents ideal coverage of subject. 334 exercises make book perfect for self-study. 55 figures. 352pp. 5⅜ × 8¼. 64025-6 Pa. $9.95

OPTICKS, Sir Isaac Newton. Newton's own experiments with spectroscopy, colors, lenses, reflection, refraction, etc., in language the layman can follow. Foreword by Albert Einstein. 532pp. 5⅜ × 8½. 60205-2 Pa. $9.95

GENERALIZED INTEGRAL TRANSFORMATIONS, A.H. Zemanian. Graduate-level study of recent generalizations of the Laplace, Mellin, Hankel, K. Weierstrass, convolution and other simple transformations. Bibliography. 320pp. 5⅜ × 8½. 65375-7 Pa. $7.95

THE ELECTROMAGNETIC FIELD, Albert Shadowitz. Comprehensive undergraduate text covers basics of electric and magnetic fields, builds up to electromagnetic theory. Also related topics, including relativity. Over 900 problems. 768pp. 5⅜ × 8¼. 65660-8 Pa. $17.95

FOURIER SERIES, Georgi P. Tolstov. Translated by Richard A. Silverman. A valuable addition to the literature on the subject, moving clearly from subject to subject and theorem to theorem. 107 problems, answers. 336pp. 5⅜ × 8½. 63317-9 Pa. $8.95

THEORY OF ELECTROMAGNETIC WAVE PROPAGATION, Charles Herach Papas. Graduate-level study discusses the Maxwell field equations, radiation from wire antennas, the Doppler effect and more. xiii + 244pp. 5⅜ × 8½. 65678-0 Pa. $6.95

DISTRIBUTION THEORY AND TRANSFORM ANALYSIS: An Introduction to Generalized Functions, with Applications, A.H. Zemanian. Provides basics of distribution theory, describes generalized Fourier and Laplace transformations. Numerous problems. 384pp. 5⅜ × 8½. 65479-6 Pa. $9.95

THE PHYSICS OF WAVES, William C. Elmore and Mark A. Heald. Unique overview of classical wave theory. Acoustics, optics, electromagnetic radiation, more. Ideal as classroom text or for self-study. Problems. 477pp. 5⅜ × 8½. 64926-1 Pa. $11.95

CALCULUS OF VARIATIONS WITH APPLICATIONS, George M. Ewing. Applications-oriented introduction to variational theory develops insight and promotes understanding of specialized books, research papers. Suitable for advanced undergraduate/graduate students as primary, supplementary text. 352pp. 5⅜ × 8½. 64856-7 Pa. $8.95

A TREATISE ON ELECTRICITY AND MAGNETISM, James Clerk Maxwell. Important foundation work of modern physics. Brings to final form Maxwell's theory of electromagnetism and rigorously derives his general equations of field theory. 1,084pp. 5⅜ × 8½. 60636-8, 60637-6 Pa., Two-vol. set $19.90

AN INTRODUCTION TO THE CALCULUS OF VARIATIONS, Charles Fox. Graduate-level text covers variations of an integral, isoperimetrical problems, least action, special relativity, approximations, more. References. 279pp. 5⅜ × 8½. 65499-0 Pa. $7.95

HYDRODYNAMIC AND HYDROMAGNETIC STABILITY, S. Chandrasekhar. Lucid examination of the Rayleigh-Benard problem; clear coverage of the theory of instabilities causing convection. 704pp. 5⅜ × 8¼. 64071-X Pa. $14.95

CALCULUS OF VARIATIONS, Robert Weinstock. Basic introduction covering isoperimetric problems, theory of elasticity, quantum mechanics, electrostatics, etc. Exercises throughout. 326pp. 5⅜ × 8½. 63069-2 Pa. $7.95

DYNAMICS OF FLUIDS IN POROUS MEDIA, Jacob Bear. For advanced students of ground water hydrology, soil mechanics and physics, drainage and irrigation engineering and more. 335 illustrations. Exercises, with answers. 784pp. 6⅛ × 9¼. 65675-6 Pa. $19.95

NUMERICAL METHODS FOR SCIENTISTS AND ENGINEERS, Richard Hamming. Classic text stresses frequency approach in coverage of algorithms, polynomial approximation, Fourier approximation, exponential approximation, other topics. Revised and enlarged 2nd edition. 721pp. 5⅜ × 8½.
65241-6 Pa. $14.95

THEORETICAL SOLID STATE PHYSICS, Vol. I: Perfect Lattices in Equilibrium; Vol. II: Non-Equilibrium and Disorder, William Jones and Norman H. March. Monumental reference work covers fundamental theory of equilibrium properties of perfect crystalline solids, non-equilibrium properties, defects and disordered systems. Appendices. Problems. Preface. Diagrams. Index. Bibliography. Total of 1,301pp. 5⅜ × 8½. Two volumes. Vol. I 65015-4 Pa. $14.95
Vol. II 65016-2 Pa. $12.95

OPTIMIZATION THEORY WITH APPLICATIONS, Donald A. Pierre. Broad-spectrum approach to important topic. Classical theory of minima and maxima, calculus of variations, simplex technique and linear programming, more. Many problems, examples. 640pp. 5⅜ × 8½.
65205-X Pa. $14.95

THE MODERN THEORY OF SOLIDS, Frederick Seitz. First inexpensive edition of classic work on theory of ionic crystals, free-electron theory of metals and semiconductors, molecular binding, much more. 736pp. 5⅜ × 8½.
65482-6 Pa. $15.95

ESSAYS ON THE THEORY OF NUMBERS, Richard Dedekind. Two classic essays by great German mathematician: on the theory of irrational numbers; and on transfinite numbers and properties of natural numbers. 115pp. 5⅜ × 8½.
21010-3 Pa. $4.95

THE FUNCTIONS OF MATHEMATICAL PHYSICS, Harry Hochstadt. Comprehensive treatment of orthogonal polynomials, hypergeometric functions, Hill's equation, much more. Bibliography. Index. 322pp. 5⅜ × 8½. 65214-9 Pa. $9.95

NUMBER THEORY AND ITS HISTORY, Oystein Ore. Unusually clear, accessible introduction covers counting, properties of numbers, prime numbers, much more. Bibliography. 380pp. 5⅜ × 8½. 65620-9 Pa. $9.95

THE VARIATIONAL PRINCIPLES OF MECHANICS, Cornelius Lanczos. Graduate level coverage of calculus of variations, equations of motion, relativistic mechanics, more. First inexpensive paperbound edition of classic treatise. Index. Bibliography. 418pp. 5⅜ × 8½. 65067-7 Pa. $10.95

MATHEMATICAL TABLES AND FORMULAS, Robert D. Carmichael and Edwin R. Smith. Logarithms, sines, tangents, trig functions, powers, roots, reciprocals, exponential and hyperbolic functions, formulas and theorems. 269pp. 5⅜ × 8½.
60111-0 Pa. $6.95

THEORETICAL PHYSICS, Georg Joos, with Ira M. Freeman. Classic overview covers essential math, mechanics, electromagnetic theory, thermodynamics, quantum mechanics, nuclear physics, other topics. First paperback edition. xxiii + 885pp. 5⅜ × 8½.
65227-0 Pa. $18.95

HANDBOOK OF MATHEMATICAL FUNCTIONS WITH FORMULAS, GRAPHS, AND MATHEMATICAL TABLES, edited by Milton Abramowitz and Irene A. Stegun. Vast compendium: 29 sets of tables, some to as high as 20 places. 1,046pp. 8 × 10½. 61272-4 Pa. $22.95

MATHEMATICAL METHODS IN PHYSICS AND ENGINEERING, John W. Dettman. Algebraically based approach to vectors, mapping, diffraction, other topics in applied math. Also generalized functions, analytic function theory, more. Exercises. 448pp. 5⅜ × 8¼. 65649-7 Pa. $9.95

A SURVEY OF NUMERICAL MATHEMATICS, David M. Young and Robert Todd Gregory. Broad self-contained coverage of computer-oriented numerical algorithms for solving various types of mathematical problems in linear algebra, ordinary and partial, differential equations, much more. Exercises. Total of 1,248pp. 5⅜ × 8½. Two volumes. Vol. I 65691-8 Pa. $14.95
Vol. II 65692-6 Pa. $14.95

TENSOR ANALYSIS FOR PHYSICISTS, J.A. Schouten. Concise exposition of the mathematical basis of tensor analysis, integrated with well-chosen physical examples of the theory. Exercises. Index. Bibliography. 289pp. 5⅜ × 8½. 65582-2 Pa. $7.95

INTRODUCTION TO NUMERICAL ANALYSIS (2nd Edition), F.B. Hildebrand. Classic, fundamental treatment covers computation, approximation, interpolation, numerical differentiation and integration, other topics. 150 new problems. 669pp. 5⅜ × 8½. 65363-3 Pa. $14.95

INVESTIGATIONS ON THE THEORY OF THE BROWNIAN MOVEMENT, Albert Einstein. Five papers (1905–8) investigating dynamics of Brownian motion and evolving elementary theory. Notes by R. Fürth. 122pp. 5⅜ × 8½. 60304-0 Pa. $4.95

NUMERICAL METHODS FOR SCIENTISTS AND ENGINEERS, Richard Hamming. Classic text stresses frequency approach in coverage of algorithms, polynomial approximation, Fourier approximation, exponential approximation, other topics. Revised and enlarged 2nd edition. 721pp. 5⅜ × 8½. 65241-6 Pa. $14.95

AN INTRODUCTION TO STATISTICAL THERMODYNAMICS, Terrell L. Hill. Excellent basic text offers wide-ranging coverage of quantum statistical mechanics, systems of interacting molecules, quantum statistics, more. 523pp. 5⅜ × 8½. 65242-4 Pa. $11.95

ELEMENTARY DIFFERENTIAL EQUATIONS, William Ted Martin and Eric Reissner. Exceptionally clear, comprehensive introduction at undergraduate level. Nature and origin of differential equations, differential equations of first, second and higher orders. Picard's Theorem, much more. Problems with solutions. 331pp. 5⅜ × 8½. 65024-3 Pa. $8.95

STATISTICAL PHYSICS, Gregory H. Wannier. Classic text combines thermodynamics, statistical mechanics and kinetic theory in one unified presentation of thermal physics. Problems with solutions. Bibliography. 532pp. 5⅜ × 8½. 65401-X Pa. $11.95

ORDINARY DIFFERENTIAL EQUATIONS, Morris Tenenbaum and Harry Pollard. Exhaustive survey of ordinary differential equations for undergraduates in mathematics, engineering, science. Thorough analysis of theorems. Diagrams. Bibliography. Index. 818pp. 5⅜ × 8½. 64940-7 Pa. $16.95

STATISTICAL MECHANICS: Principles and Applications, Terrell L. Hill. Standard text covers fundamentals of statistical mechanics, applications to fluctuation theory, imperfect gases, distribution functions, more. 448pp. 5⅜ × 8½. 65390-0 Pa. $9.95

ORDINARY DIFFERENTIAL EQUATIONS AND STABILITY THEORY: An Introduction, David A. Sánchez. Brief, modern treatment. Linear equation, stability theory for autonomous and nonautonomous systems, etc. 164pp. 5⅜ × 8¼. 63828-6 Pa. $5.95

THIRTY YEARS THAT SHOOK PHYSICS: The Story of Quantum Theory, George Gamow. Lucid, accessible introduction to influential theory of energy and matter. Careful explanations of Dirac's anti-particles, Bohr's model of the atom, much more. 12 plates. Numerous drawings. 240pp. 5⅜ × 8½. 24895-X Pa. $6.95

THEORY OF MATRICES, Sam Perlis. Outstanding text covering rank, non-singularity and inverses in connection with the development of canonical matrices under the relation of equivalence, and without the intervention of determinants. Includes exercises. 237pp. 5⅜ × 8½. 66810-X Pa. $7.95

GREAT EXPERIMENTS IN PHYSICS: Firsthand Accounts from Galileo to Einstein, edited by Morris H. Shamos. 25 crucial discoveries: Newton's laws of motion, Chadwick's study of the neutron, Hertz on electromagnetic waves, more. Original accounts clearly annotated. 370pp. 5⅜ × 8½. 25346-5 Pa. $9.95

INTRODUCTION TO PARTIAL DIFFERENTIAL EQUATIONS WITH AP-PLICATIONS, E.C. Zachmanoglou and Dale W. Thoe. Essentials of partial differential equations applied to common problems in engineering and the physical sciences. Problems and answers. 416pp. 5⅜ × 8½. 65251-3 Pa. $10.95

BURNHAM'S CELESTIAL HANDBOOK, Robert Burnham, Jr. Thorough guide to the stars beyond our solar system. Exhaustive treatment. Alphabetical by constellation: Andromeda to Cetus in Vol. 1; Chamaeleon to Orion in Vol. 2; and Pavo to Vulpecula in Vol. 3. Hundreds of illustrations. Index in Vol. 3. 2,000pp. 6⅛ × 9¼. 23567-X, 23568-8, 23673-0 Pa., Three-vol. set $41.85

ASYMPTOTIC EXPANSIONS FOR ORDINARY DIFFERENTIAL EQUA-TIONS, Wolfgang Wasow. Outstanding text covers asymptotic power series, Jordan's canonical form, turning point problems, singular perturbations, much more. Problems. 384pp. 5⅜ × 8½. 65456-7 Pa. $9.95

AMATEUR ASTRONOMER'S HANDBOOK, J.B. Sidgwick. Timeless, comprehensive coverage of telescopes, mirrors, lenses, mountings, telescope drives, micrometers, spectroscopes, more. 189 illustrations. 576pp. 5⅜ × 8¼. (USO) 24034-7 Pa. $9.95

SPECIAL FUNCTIONS, N.N. Lebedev. Translated by Richard Silverman. Famous Russian work treating more important special functions, with applications to specific problems of physics and engineering. 38 figures. 308pp. 5⅜ × 8½.
60624-4 Pa. $7.95

OBSERVATIONAL ASTRONOMY FOR AMATEURS, J.B. Sidgwick. Mine of useful data for observation of sun, moon, planets, asteroids, aurorae, meteors, comets, variables, binaries, etc. 39 illustrations. 384pp. 5⅜ × 8¼. (Available in U.S. only)
24033-9 Pa. $8.95

INTEGRAL EQUATIONS, F.G. Tricomi. Authoritative, well-written treatment of extremely useful mathematical tool with wide applications. Volterra Equations, Fredholm Equations, much more. Advanced undergraduate to graduate level. Exercises. Bibliography. 238pp. 5⅜ × 8½.
64828-1 Pa. $7.95

CELESTIAL OBJECTS FOR COMMON TELESCOPES, T.W. Webb. Inestimable aid for locating and identifying nearly 4,000 celestial objects. 77 illustrations. 645pp. 5⅜ × 8½.
20917-2, 20918-0 Pa., Two-vol. set $12.00

MODERN NONLINEAR EQUATIONS, Thomas L. Saaty. Emphasizes practical solution of problems; covers seven types of equations. ". . . a welcome contribution to the existing literature. . . ."—*Math Reviews.* 490pp. 5⅜ × 8½. 64232-1 Pa. $9.95

FUNDAMENTALS OF ASTRODYNAMICS, Roger Bate et al. Modern approach developed by U.S. Air Force Academy. Designed as a first course. Problems, exercises. Numerous illustrations. 455pp. 5⅜ × 8½.
60061-0 Pa. $8.95

INTRODUCTION TO LINEAR ALGEBRA AND DIFFERENTIAL EQUATIONS, John W. Dettman. Excellent text covers complex numbers, determinants, orthonormal bases, Laplace transforms, much more. Exercises with solutions. Undergraduate level. 416pp. 5⅜ × 8½.
65191-6 Pa. $9.95

INCOMPRESSIBLE AERODYNAMICS, edited by Bryan Thwaites. Covers theoretical and experimental treatment of the uniform flow of air and viscous fluids past two-dimensional aerofoils and three-dimensional wings; many other topics. 654pp. 5⅜ × 8½.
65465-6 Pa. $16.95

INTRODUCTION TO DIFFERENCE EQUATIONS, Samuel Goldberg. Exceptionally clear exposition of important discipline with applications to sociology, psychology, economics. Many illustrative examples; over 250 problems. 260pp. 5⅜ × 8½.
65084-7 Pa. $7.95

LAMINAR BOUNDARY LAYERS, edited by L. Rosenhead. Engineering classic covers steady boundary layers in two- and three-dimensional flow, unsteady boundary layers, stability, observational techniques, much more. 708pp. 5⅜ × 8½.
65646-2 Pa. $15.95

LECTURES ON CLASSICAL DIFFERENTIAL GEOMETRY, Second Edition, Dirk J. Struik. Excellent brief introduction covers curves, theory of surfaces, fundamental equations, geometry on a surface, conformal mapping, other topics. Problems. 240pp. 5⅜ × 8½.
65609-8 Pa. $7.95

ROTARY-WING AERODYNAMICS, W.Z. Stepniewski. Clear, concise text covers aerodynamic phenomena of the rotor and offers guidelines for helicopter performance evaluation. Originally prepared for NASA. 537 figures. 640pp. 6⅛ × 9¼.
64647-5 Pa. $15.95

DIFFERENTIAL GEOMETRY, Heinrich W. Guggenheimer. Local differential geometry as an application of advanced calculus and linear algebra. Curvature, transformation groups, surfaces, more. Exercises. 62 figures. 378pp. 5⅜ × 8½.
63433-7 Pa. $7.95

INTRODUCTION TO SPACE DYNAMICS, William Tyrrell Thomson. Comprehensive, classic introduction to space-flight engineering for advanced undergraduate and graduate students. Includes vector algebra, kinematics, transformation of coordinates. Bibliography. Index. 352pp. 5⅜ × 8½.   65113-4 Pa. $8.95

A SURVEY OF MINIMAL SURFACES, Robert Osserman. Up-to-date, in-depth discussion of the field for advanced students. Corrected and enlarged edition covers new developments. Includes numerous problems. 192pp. 5⅜ × 8½.
64998-9 Pa. $8.95

ANALYTICAL MECHANICS OF GEARS, Earle Buckingham. Indispensable reference for modern gear manufacture covers conjugate gear-tooth action, gear-tooth profiles of various gears, many other topics. 263 figures. 102 tables. 546pp. 5⅜ × 8½.   65712-4 Pa. $11.95

SET THEORY AND LOGIC, Robert R. Stoll. Lucid introduction to unified theory of mathematical concepts. Set theory and logic seen as tools for conceptual understanding of real number system. 496pp. 5⅜ × 8¼.   63829-4 Pa. $10.95

A HISTORY OF MECHANICS, René Dugas. Monumental study of mechanical principles from antiquity to quantum mechanics. Contributions of ancient Greeks, Galileo, Leonardo, Kepler, Lagrange, many others. 671pp. 5⅜ × 8½.
65632-2 Pa. $14.95

FAMOUS PROBLEMS OF GEOMETRY AND HOW TO SOLVE THEM, Benjamin Bold. Squaring the circle, trisecting the angle, duplicating the cube: learn their history, why they are impossible to solve, then solve them yourself. 128pp. 5⅜ × 8½.   24297-8 Pa. $3.95

MECHANICAL VIBRATIONS, J.P. Den Hartog. Classic textbook offers lucid explanations and illustrative models, applying theories of vibrations to a variety of practical industrial engineering problems. Numerous figures. 233 problems, solutions. Appendix. Index. Preface. 436pp. 5⅜ × 8½.   64785-4 Pa. $9.95

CURVATURE AND HOMOLOGY, Samuel I. Goldberg. Thorough treatment of specialized branch of differential geometry. Covers Riemannian manifolds, topology of differentiable manifolds, compact Lie groups, other topics. Exercises. 315pp. 5⅜ × 8½.   64314-X Pa. $8.95

HISTORY OF STRENGTH OF MATERIALS, Stephen P. Timoshenko. Excellent historical survey of the strength of materials with many references to the theories of elasticity and structure. 245 figures. 452pp. 5⅜ × 8½. 61187-6 Pa. $10.95

GEOMETRY OF COMPLEX NUMBERS, Hans Schwerdtfeger. Illuminating, widely praised book on analytic geometry of circles, the Moebius transformation, and two-dimensional non-Euclidean geometries. 200pp. 5⅜ × 8¼. 63830-8 Pa. $6.95

MECHANICS, J.P. Den Hartog. A classic introductory text or refresher. Hundreds of applications and design problems illuminate fundamentals of trusses, loaded beams and cables, etc. 334 answered problems. 462pp. 5⅜ × 8½. 60754-2 Pa. $8.95

TOPOLOGY, John G. Hocking and Gail S. Young. Superb one-year course in classical topology. Topological spaces and functions, point-set topology, much more. Examples and problems. Bibliography. Index. 384pp. 5⅜ × 8¼. 65676-4 Pa. $8.95

STRENGTH OF MATERIALS, J.P. Den Hartog. Full, clear treatment of basic material (tension, torsion, bending, etc.) plus advanced material on engineering methods, applications. 350 answered problems. 323pp. 5⅜ × 8½. 60755-0 Pa. $8.95

ELEMENTARY CONCEPTS OF TOPOLOGY, Paul Alexandroff. Elegant, intuitive approach to topology from set-theoretic topology to Betti groups; how concepts of topology are useful in math and physics. 25 figures. 57pp. 5⅜ × 8½. 60747-X Pa. $3.50

ADVANCED STRENGTH OF MATERIALS, J.P. Den Hartog. Superbly written advanced text covers torsion, rotating disks, membrane stresses in shells, much more. Many problems and answers. 388pp. 5⅜ × 8½. 65407-9 Pa. $9.95

COMPUTABILITY AND UNSOLVABILITY, Martin Davis. Classic graduate-level introduction to theory of computability, usually referred to as theory of recurrent functions. New preface and appendix. 288pp. 5⅜ × 8½. 61471-9 Pa. $7.95

GENERAL CHEMISTRY, Linus Pauling. Revised 3rd edition of classic first-year text by Nobel laureate. Atomic and molecular structure, quantum mechanics, statistical mechanics, thermodynamics correlated with descriptive chemistry. Problems. 992pp. 5⅜ × 8½. 65622-5 Pa. $19.95

AN INTRODUCTION TO MATRICES, SETS AND GROUPS FOR SCIENCE STUDENTS, G. Stephenson. Concise, readable text introduces sets, groups, and most importantly, matrices to undergraduate students of physics, chemistry, and engineering. Problems. 164pp. 5⅜ × 8½. 65077-4 Pa. $6.95

THE HISTORICAL BACKGROUND OF CHEMISTRY, Henry M. Leicester. Evolution of ideas, not individual biography. Concentrates on formulation of a coherent set of chemical laws. 260pp. 5⅜ × 8½. 61053-5 Pa. $6.95

THE PHILOSOPHY OF MATHEMATICS: An Introductory Essay, Stephan Körner. Surveys the views of Plato, Aristotle, Leibniz & Kant concerning propositions and theories of applied and pure mathematics. Introduction. Two appendices. Index. 198pp. 5⅜ × 8½. 25048-2 Pa. $6.95

THE DEVELOPMENT OF MODERN CHEMISTRY, Aaron J. Ihde. Authoritative history of chemistry from ancient Greek theory to 20th-century innovation. Covers major chemists and their discoveries. 209 illustrations. 14 tables. Bibliographies. Indices. Appendices. 851pp. 5⅜ × 8½. 64235-6 Pa. $17.95

DE RE METALLICA, Georgius Agricola. The famous Hoover translation of greatest treatise on technological chemistry, engineering, geology, mining of early modern times (1556). All 289 original woodcuts. 638pp. 6¾ × 11.
60006-8 Pa. $17.95

SOME THEORY OF SAMPLING, William Edwards Deming. Analysis of the problems, theory and design of sampling techniques for social scientists, industrial managers and others who find statistics increasingly important in their work. 61 tables. 90 figures. xvii + 602pp. 5⅜ × 8½.
64684-X Pa. $15.95

THE VARIOUS AND INGENIOUS MACHINES OF AGOSTINO RAMELLI: A Classic Sixteenth-Century Illustrated Treatise on Technology, Agostino Ramelli. One of the most widely known and copied works on machinery in the 16th century. 194 detailed plates of water pumps, grain mills, cranes, more. 608pp. 9 × 12. (EBE)
25497-6 Clothbd. $34.95

LINEAR PROGRAMMING AND ECONOMIC ANALYSIS, Robert Dorfman, Paul A. Samuelson and Robert M. Solow. First comprehensive treatment of linear programming in standard economic analysis. Game theory, modern welfare economics, Leontief input-output, more. 525pp. 5⅜ × 8½.
65491-5 Pa. $13.95

ELEMENTARY DECISION THEORY, Herman Chernoff and Lincoln E. Moses. Clear introduction to statistics and statistical theory covers data processing, probability and random variables, testing hypotheses, much more. Exercises. 364pp. 5⅜ × 8½.
65218-1 Pa. $9.95

THE COMPLEAT STRATEGYST: Being a Primer on the Theory of Games of Strategy, J.D. Williams. Highly entertaining classic describes, with many illustrated examples, how to select best strategies in conflict situations. Prefaces. Appendices. 268pp. 5⅜ × 8½.
25101-2 Pa. $6.95

MATHEMATICAL METHODS OF OPERATIONS RESEARCH, Thomas L. Saaty. Classic graduate-level text covers historical background, classical methods of forming models, optimization, game theory, probability, queueing theory, much more. Exercises. Bibliography. 448pp. 5⅜ × 8¼.
65703-5 Pa. $12.95

CONSTRUCTIONS AND COMBINATORIAL PROBLEMS IN DESIGN OF EXPERIMENTS, Damaraju Raghavarao. In-depth reference work examines orthogonal Latin squares, incomplete block designs, tactical configuration, partial geometry, much more. Abundant explanations, examples. 416pp. 5⅜ × 8¼.
65685-3 Pa. $10.95

THE ABSOLUTE DIFFERENTIAL CALCULUS (CALCULUS OF TENSORS), Tullio Levi-Civita. Great 20th-century mathematician's classic work on material necessary for mathematical grasp of theory of relativity. 452pp. 5⅜ × 8½.
63401-9 Pa. $9.95

VECTOR AND TENSOR ANALYSIS WITH APPLICATIONS, A.I. Borisenko and I.E. Tarapov. Concise introduction. Worked-out problems, solutions, exercises. 257pp. 5⅜ × 8¼.
63833-2 Pa. $6.95

THE FOUR-COLOR PROBLEM: Assaults and Conquest, Thomas L. Saaty and Paul G. Kainen. Engrossing, comprehensive account of the century-old combinatorial topological problem, its history and solution. Bibliographies. Index. 110 figures. 228pp. 5⅜ × 8½. 65092-8 Pa. $6.95

CATALYSIS IN CHEMISTRY AND ENZYMOLOGY, William P. Jencks. Exceptionally clear coverage of mechanisms for catalysis, forces in aqueous solution, carbonyl- and acyl-group reactions, practical kinetics, more. 864pp. 5⅜ × 8½. 65460-5 Pa. $19.95

PROBABILITY: An Introduction, Samuel Goldberg. Excellent basic text covers set theory, probability theory for finite sample spaces, binomial theorem, much more. 360 problems. Bibliographies. 322pp. 5⅜ × 8½. 65252-1 Pa. $8.95

LIGHTNING, Martin A. Uman. Revised, updated edition of classic work on the physics of lightning. Phenomena, terminology, measurement, photography, spectroscopy, thunder, more. Reviews recent research. Bibliography. Indices. 320pp. 5⅜ × 8¼. 64575-4 Pa. $8.95

PROBABILITY THEORY: A Concise Course, Y.A. Rozanov. Highly readable, self-contained introduction covers combination of events, dependent events, Bernoulli trials, etc. Translation by Richard Silverman. 148pp. 5⅜ × 8¼. 63544-9 Pa. $5.95

THE CEASELESS WIND: An Introduction to the Theory of Atmospheric Motion, John A. Dutton. Acclaimed text integrates disciplines of mathematics and physics for full understanding of dynamics of atmospheric motion. Over 400 problems. Index. 97 illustrations. 640pp. 6 × 9. 65096-0 Pa. $17.95

STATISTICS MANUAL, Edwin L. Crow, et al. Comprehensive, practical collection of classical and modern methods prepared by U.S. Naval Ordnance Test Station. Stress on use. Basics of statistics assumed. 288pp. 5⅜ × 8½. 60599-X Pa. $6.95

DICTIONARY/OUTLINE OF BASIC STATISTICS, John E. Freund and Frank J. Williams. A clear concise dictionary of over 1,000 statistical terms and an outline of statistical formulas covering probability, nonparametric tests, much more. 208pp. 5⅜ × 8½. 66796-0 Pa. $6.95

STATISTICAL METHOD FROM THE VIEWPOINT OF QUALITY CONTROL, Walter A. Shewhart. Important text explains regulation of variables, uses of statistical control to achieve quality control in industry, agriculture, other areas. 192pp. 5⅜ × 8½. 65232-7 Pa. $7.95

THE INTERPRETATION OF GEOLOGICAL PHASE DIAGRAMS, Ernest G. Ehlers. Clear, concise text emphasizes diagrams of systems under fluid or containing pressure; also coverage of complex binary systems, hydrothermal melting, more. 288pp. 6½ × 9¼. 65389-7 Pa. $10.95

STATISTICAL ADJUSTMENT OF DATA, W. Edwards Deming. Introduction to basic concepts of statistics, curve fitting, least squares solution, conditions without parameter, conditions containing parameters. 26 exercises worked out. 271pp. 5⅜ × 8½. 64685-8 Pa. $8.95

TENSOR CALCULUS, J.L. Synge and A. Schild. Widely used introductory text covers spaces and tensors, basic operations in Riemannian space, non-Riemannian spaces, etc. 324pp. 5⅜ × 8¼.   63612-7 Pa. $7.95

A CONCISE HISTORY OF MATHEMATICS, Dirk J. Struik. The best brief history of mathematics. Stresses origins and covers every major figure from ancient Near East to 19th century. 41 illustrations. 195pp. 5⅜ × 8½.   60255-9 Pa. $7.95

A SHORT ACCOUNT OF THE HISTORY OF MATHEMATICS, W.W. Rouse Ball. One of clearest, most authoritative surveys from the Egyptians and Phoenicians through 19th-century figures such as Grassman, Galois, Riemann. Fourth edition. 522pp. 5⅜ × 8½.   20630-0 Pa. $10.95

HISTORY OF MATHEMATICS, David E. Smith. Nontechnical survey from ancient Greece and Orient to late 19th century; evolution of arithmetic, geometry, trigonometry, calculating devices, algebra, the calculus. 362 illustrations. 1,355pp. 5⅜ × 8½.   20429-4, 20430-8 Pa., Two-vol. set $23.90

THE GEOMETRY OF RENÉ DESCARTES, René Descartes. The great work founded analytical geometry. Original French text, Descartes' own diagrams, together with definitive Smith-Latham translation. 244pp. 5⅜ × 8½.
                                        60068-8 Pa. $6.95

THE ORIGINS OF THE INFINITESIMAL CALCULUS, Margaret E. Baron. Only fully detailed and documented account of crucial discipline: origins; development by Galileo, Kepler, Cavalieri; contributions of Newton, Leibniz, more. 304pp. 5⅜ × 8½. (Available in U.S. and Canada only)   65371-4 Pa. $9.95

THE HISTORY OF THE CALCULUS AND ITS CONCEPTUAL DEVELOPMENT, Carl B. Boyer. Origins in antiquity, medieval contributions, work of Newton, Leibniz, rigorous formulation. Treatment is verbal. 346pp. 5⅜ × 8½.
                                        60509-4 Pa. $7.95

THE THIRTEEN BOOKS OF EUCLID'S ELEMENTS, translated with introduction and commentary by Sir Thomas L. Heath. Definitive edition. Textual and linguistic notes, mathematical analysis. 2,500 years of critical commentary. Not abridged. 1,414pp. 5⅜ × 8½.   60088-2, 60089-0, 60090-4 Pa., Three-vol. set $29.85

GAMES AND DECISIONS: Introduction and Critical Survey, R. Duncan Luce and Howard Raiffa. Superb nontechnical introduction to game theory, primarily applied to social sciences. Utility theory, zero-sum games, n-person games, decision-making, much more. Bibliography. 509pp. 5⅜ × 8½.   65943-7 Pa. $11.95

THE HISTORICAL ROOTS OF ELEMENTARY MATHEMATICS, Lucas N.H. Bunt, Phillip S. Jones, and Jack D. Bedient. Fundamental underpinnings of modern arithmetic, algebra, geometry and number systems derived from ancient civilizations. 320pp. 5⅜ × 8½.   25563-8 Pa. $8.95

CALCULUS REFRESHER FOR TECHNICAL PEOPLE, A. Albert Klaf. Covers important aspects of integral and differential calculus via 756 questions. 566 problems, most answered. 431pp. 5⅜ × 8½.   20370-0 Pa. $8.95

CATALOG OF DOVER BOOKS

CHALLENGING MATHEMATICAL PROBLEMS WITH ELEMENTARY SOLUTIONS, A.M. Yaglom and I.M. Yaglom. Over 170 challenging problems on probability theory, combinatorial analysis, points and lines, topology, convex polygons, many other topics. Solutions. Total of 445pp. 5⅜ × 8½. Two-vol. set.
Vol. I 65536-9 Pa. $6.95
Vol. II 65537-7 Pa. $6.95

FIFTY CHALLENGING PROBLEMS IN PROBABILITY WITH SOLUTIONS, Frederick Mosteller. Remarkable puzzlers, graded in difficulty, illustrate elementary and advanced aspects of probability. Detailed solutions. 88pp. 5⅜ × 8½.
65355-2 Pa. $4.95

EXPERIMENTS IN TOPOLOGY, Stephen Barr. Classic, lively explanation of one of the byways of mathematics. Klein bottles, Moebius strips, projective planes, map coloring, problem of the Koenigsberg bridges, much more, described with clarity and wit. 43 figures. 210pp. 5⅜ × 8½.
25933-1 Pa. $5.95

RELATIVITY IN ILLUSTRATIONS, Jacob T. Schwartz. Clear nontechnical treatment makes relativity more accessible than ever before. Over 60 drawings illustrate concepts more clearly than text alone. Only high school geometry needed. Bibliography. 128pp. 6⅛ × 9¼.
25965-X Pa. $6.95

AN INTRODUCTION TO ORDINARY DIFFERENTIAL EQUATIONS, Earl A. Coddington. A thorough and systematic first course in elementary differential equations for undergraduates in mathematics and science, with many exercises and problems (with answers). Index. 304pp. 5⅜ × 8½.
65942-9 Pa. $8.95

FOURIER SERIES AND ORTHOGONAL FUNCTIONS, Harry F. Davis. An incisive text combining theory and practical example to introduce Fourier series, orthogonal functions and applications of the Fourier method to boundary-value problems. 570 exercises. Answers and notes. 416pp. 5⅜ × 8½.
65973-9 Pa. $9.95

THE THEORY OF BRANCHING PROCESSES, Theodore E. Harris. First systematic, comprehensive treatment of branching (i.e. multiplicative) processes and their applications. Galton-Watson model, Markov branching processes, electron-photon cascade, many other topics. Rigorous proofs. Bibliography. 240pp. 5⅜ × 8½.
65952-6 Pa. $6.95

AN INTRODUCTION TO ALGEBRAIC STRUCTURES, Joseph Landin. Superb self-contained text covers "abstract algebra": sets and numbers, theory of groups, theory of rings, much more. Numerous well-chosen examples, exercises. 247pp. 5⅜ × 8½.
65940-2 Pa. $6.95

Prices subject to change without notice.
Available at your book dealer or write for free Mathematics and Science Catalog to Dept. GI, Dover Publications, Inc., 31 East 2nd St., Mineola, N.Y. 11501. Dover publishes more than 175 books each year on science, elementary and advanced mathematics, biology, music, art, literature, history, social sciences and other areas.

Abacist vs. Algorismist

From Gregor Reisch: Margarita Philosophica
Strassbourg 1504